Springer Theses

Recognizing Outstanding Ph.D. Research

For further volumes:
http://www.springer.com/series/8790

Aims and Scope

The series "Springer Theses" brings together a selection of the very best Ph.D. theses from around the world and across the physical sciences. Nominated and endorsed by two recognized specialists, each published volume has been selected for its scientific excellence and the high impact of its contents for the pertinent field of research. For greater accessibility to non-specialists, the published versions include an extended introduction, as well as a foreword by the student's supervisor explaining the special relevance of the work for the field. As a whole, the series will provide a valuable resource both for newcomers to the research fields described, and for other scientists seeking detailed background information on special questions. Finally, it provides an accredited documentation of the valuable contributions made by today's younger generation of scientists.

Theses are accepted into the series by invited nomination only and must fulfill all of the following criteria

- They must be written in good English.
- The topic of should fall within the confines of Chemistry, Physics and related interdisciplinary fields such as Materials, Nanoscience, Chemical Engineering, Complex Systems and Biophysics.
- The work reported in the thesis must represent a significant scientific advance.
- If the thesis includes previously published material, permission to reproduce this must be gained from the respective copyright holder.
- They must have been examined and passed during the 12 months prior to nomination.
- Each thesis should include a foreword by the supervisor outlining the significance of its content.
- The theses should have a clearly defined structure including an introduction accessible to scientists not expert in that particular field.

Boyang Liu

Muonium–Antimuonium Oscillations in an Extended Minimal Supersymmetric Standard Model

Doctoral Thesis accepted by Purdue University, West Lafayette, USA

Springer

Dr. Boyang Liu
Chinese Academy of Sciences
Institute of Physics
P.O. Box 603
100190 Beijing
People's Republic of China
e-mail: boyangleo@gmail.com

Supervisor
Dr. Sherwin T. Love
Department of Physics
Purdue University
1396 Physics Building
West Lafayette, IN 47907
USA
e-mail: love@physics.purdue.edu

ISSN 2190-5053 ISSN 2190-5061 (eBook)

ISBN 978-1-4614-2805-3 ISBN 978-1-4419-8330-5 (eBook)

DOI 10.1007/978-1-4419-8330-5

Springer New York Dordrecht Heidelberg London

Softcover re-print of the Hardcover 1st edition 2011

Cover design: eStudio Calamar, Berlin/Figueres

Printed on acid-free paper

Springer is part of Springer Science+Business Media (www.springer.com)

Supervisor's Foreword

The results of a variety of neutrino oscillation experiments involving solar, atmospheric, accelerator and reactor neutrinos and anti-neutrinos have led to the conclusion that neutrinos produced in a well defined flavor eigenstate can be detected in a different flavor eigenstate after propagating a macroscopic distance. The simplest interpretation of this oscillation phenomenon is that the neutrinos have finite masses and that the neutrinos flavor eigenstates differ from their mass eigenstates. That is, the neutrino flavors mix. The existence of finite neutrino masses qualifies as the first empirical evidence for physics beyond the minimal Standard Model of particle physics. At present, there are still many questions regarding the nature of the neutrino masses which remain unanswered. For example, are neutrinos Dirac or Majorana fermions? It follows that the answers to neutrino related issues should add to the knowledge about the precise nature of the interactions which exist beyond the Standard Model. The existence of massive neutrinos opens the possibility of muonium–antimuonium oscillations provided the neutrinos are Majorana fermions. This process has been the subject of experimental search at the Paul Scherrer Institut (PSI) in Switzerland for a number of years. Thus far, only a lower bound has been placed on the muonium–antimuonium oscillation time scale. The PhD thesis of Boyang Liu involves the computation of the muonium–antimuonium oscillation time scale in a variety of Standard Model extensions. First he demonstrated its gauge invariance in the Standard Model modified only with the inclusion of right-handed neutrinos which were used to generate light neutrino masses via the see-saw mechanism. He explicitly displayed the gauge independence of the various 1-loop contributions. Next he considered the calculation of the muonium–antimuonium oscillation time scale in a supersymmetric (SUSY) extension of the Standard Model where there exist additional intermediate states which can contribute to the oscillation process thus possibly enhancing the rate to a range which is at or near to the current experimental bound. This thesis considers the Minimal Supersymmetric Standard Model (MSSM) modified only by the inclusion of right handed neutrino super-fields while allowing for intra-generational lepton number violation. Using the current experimental limits on the SUSY partner masses, he found that the computed oscillation time scale could indeed be lowered but it is still, in general, a couple of orders of magnitude above the current limit. Only in the special case when the mixing between the two Higgs doublets of the MSSM given by tan β is very small is the oscillation process sufficiently enhanced. In that case, the current experimental limit on the muonium–antimunium conversion can be used to bound tan $\beta > 10^{-7}$.

October 2010

Sherwin T. Love

Preface

The electron and muon number violating muonium–antimuonium oscillation process in two different models is investigated. First, modifying the Standard Model only by the inclusion of singlet right-handed neutrinos and allowing for general renormalizable interactions producing neutrino masses and mixing, the leading order matrix element contribution to this process is computed in R_ξ gauge thereby establishing the gauge invariance to this order. To give a natural explanation of the smallness of the observed neutrino masses, the see-saw mechanism is explored resulting in three light Majorana neutrinos and three heavy Majorana neutrinos with mass scale $M_R \gg M_W$. Present experimental limits set by the nonobservation of the oscillation process sets a lower limit on M_R of roughly of order 600 GeV. Second, modifying the Minimal Supersymmetric Standard Model by the inclusion of three right-handed neutrino superfields and allowing only intra-generation lepton number violation but not inter-generation lepton number mixing, the muonium–antimuonium conversion can occur while the process $\mu \to e\gamma$ is forbidden. For a wide range of the parameters, the contributions to the muonium–antimuonium oscillation time scale are at least two orders of magnitude below the sensitivity of current experiments. However, if the ratio of the two Higgs field VEVs, tan β, is very small, there is a limited possibility that the contributions are large enough for the present experimental limit to provide an inequality relating tan β with the light neutrino mass scale m_ν which is generated by see-saw mechanism. The resultant lower bound on tan β as a function of m_ν is more stringent than the analogous bounds arising from the muon and electron anomalous magnetic moments as computed using this model.

Beijing, May 2010 Boyang Liu

Acknowledgments

I thank all the individuals that have been helpful and supportive to me both professionally and personally in the completion of this work and my life.

I thank my major Professor, Sherwin Love, for his ideas, suggestions and profound professional insight. Everything that I learned from him has built a very solid scientific background for me. He is also such a nice person. I still remember that he helped me to revise my papers and postdoc application documents word by word. I really appreciate the time he shared with me.

Many thanks to Professor Thomas Clark, Professor Tzee-Ke Kuo and Dr. Chi Xiong for all the valuable discussions with them.

My thanks to my parents for supporting me in all my pursuits. They always encourage me to follow my heart and do the things I am really interested in. This precious love is an irreplaceable contribution to the completion of this thesis.

Last, but by no means least, thanks to all my friends. Without them I would have such an enjoyable life in Purdue.

I have probably left out many others, who deserve being mentioned here, to whom I apologize for the lack of space and my momentary lapse of memory.

Contents

Chapter 1
Introduction to Muonium–Antimuonium Oscillation

The time-dependent oscillation between two distinct levels or particle species is an interesting quantum mechanical phenomenon which has been widely studied in many physical systems varying from a particle moving in a double-well potential of the ammonia molecule to oscillations in the neutral $K^0 - \bar{K}^0$ and $B^0 - \bar{B}^0$ meson systems.[1] It was suggested roughly 60 years ago (Pontecorvo 1957) that there may be a spontaneous conversion between muonium and antimuonium resulting in an associated oscillation effect. Muonium (M) is the Coulombic bound state of an electron and an antimuon ($e^-\mu^+$), while antimuonium ($\bar{M}$) is the Coulombic bound state of a positron and a muon ($e^+\mu^-$). Since it has no hadronic constituents, muonium is an ideal place to test electroweak interactions. Of particular interest is that such a muonium–antimuonium oscillation is totally forbidden within the Standard Model because the process violates the individual electron and muon number conservation laws by two units. Hence, its observation will be a clear signal of physics beyond the Standard Model. Since the initial suggestion, experimental searches have been conducted (Huber 1990; Matthias et al. 1991; Abela et al. 1996; Willmann et al. 1999) and a variety of theoretical models have been proposed which can give rise to such a muonium–antimuonium conversion. These include interactions which can be mediated by (a) a doubly charged Higgs boson Δ^{++} (Halprin 1982; Herczeg and Masiero 1992), which is contained in a left–right symmetric model, (b) massive Majorana neutrinos (Clark and Love 2004; Cvetic et al. 2005; Liu 2008), or (c) the τ-sneutrino in an R-parity violation supersymmetric model (Halprin and Masiero 1993).

[1] For a general review of present state of oscillation phenomena in neutral meson systems, see, for example Battaglia et al. (2002). For a general discussion of mixing in the neutral $K^0 - \bar{K}^0$ and $B^0 - \bar{B}^0$, see, for example Buras (2000). Analogous oscillation phenomena has also been speculated upon for the neutron–antineutron system. Here the mixing matrix elements arise from an underlying baryon number violating grand unified theory. For various discussions, see Glashow (1979); Marshak and Mohapatra (1980) and Kuo and Love (1980).

B. Liu, *Muonium–Antimuonium Oscillations in an Extended Minimal Supersymmetric Standard Model*, Springer Theses,
DOI: 10.1007/978-1-4419-8330-5_1,

In Chap. 2, we focus on the muonium–antimuonium oscillations in a modified Standard Model which includes singlet right-handed neutrinos. There is now compelling evidence of the existence of neutrino oscillations from the experimental study of atmospheric and solar neutrinos (Fukuda et al. 1988; Ahmad et al. 2001, 2002; Eguchi et al. 2003; Ahn et al. 2003). That implies nonzero neutrino masses and mixing matrix elements. The size and nature of the neutrino mass and the associated mixing is still an open question subject to experimental determination and theoretical speculation (Altarelli and Feruglio 2002; Fogli et al. 2002; Gonzalez-Garcia and Nir 2003). One simple neutrino mass model is obtained by modifying the Standard Model by including singlet right-handed neutrinos and allowing for a general mass matrix for neutrinos. Left-handed neutrinos along with their charged leptonic partners are components of $SU(2)_L$ doublets and experience the weak interaction while any right-handed neutrinos are completely neutral under the Standard Model gauge group. The see-saw mechanism (Minkowski 1977; Yanagida 1979; Gell-Mann et al. 1979; Glashow 1980; Mohapatra and Senjanovic 1980) provides a natural explanation of the smallness of the three light Majorana neutrino masses, while ensuring that the other three Majorana neutrinos are heavy. Such a model could also lead to the muonium–antimuonium oscillation process. In order for there to be a nontrivial mixing between muonium and antimuonium, the individual electron and muon number conservation must be violated. Such a situation results provided the neutrinos are massive particles which mix amongst the various generations. This criterion can be met by the modified Standard Model and the $e^-\mu^+$ and $e^+\mu^-$ states could indeed mix. In Chaps. 3 and 4 we consider the muonium–antimuonium oscillation process in the Minimal Supersymmetric Standard Model (MSSM) extended by the inclusion of three right-handed neutrino superfields and where lepton flavor mixing is absent. This assumption automatically forbids the $\mu \to e\gamma$ decay, so that the muonium–antimuonium oscillation process can be used to constrain the model parameters. In order for there to be a nontrivial mixing between the muonium and antimuonium, the individual electron and muon number conservation must be violated by two units. Such a situation will result provided that the neutrinos are massive Majorana particles or the mass diagonal sneutrinos are lepton number violating scalar particles. The present experimental limit (Willmann et al. 1999) on the non-observation of muonium–antimuonium oscillation translates into a bound on the coupling constant of the effective Lagrangian of this oscillation process. Calculations show that for a wide range of the parameters, the contributions to the muonium–antimuonium oscillation time scale are at least two orders of magnitude below the sensitivity of current experiments. However, if the ratio of the two Higgs field VEVs, $\tan\beta$, is very small, there is a limited possibility that the contributions are large enough for the present experimental limit to provide an inequality relating $\tan\beta$ with the light neutrino mass scale m_ν. In Chap. 5 the corrections of muon and electron anomalous magnetic moments in the Extended Minimal Supersymmetric Standard Model (EMSSM) are calculated. Then two lower bounds on $\tan\beta$ as a function of the light neutrino mass scale m_μ are generated by the measurements of the muon and electron anomalous magnetic moments, respectively. Comparison shows that the bound from muonium–antimuonium oscillation is more stringent than ones from muon and electron anomalous magnetic moments. Finally, Chap. 6 presents the conclusions.

References

R. Abela et al., Phys. Rev. Lett. **77**, 1950 (1996)

Q.R. Ahmad et al., Phys. Rev. Lett. **87**, 071301 (2001), nucl-ex/0106015

Q.R. Ahmad et al., Phys. Rev. Lett. **89**, 011301 (2002), nucl-ex/0204008

M.H. Ahn et al., Phys. Rev. Lett. **90**, 041801 (2003), hep-ex/0212007

G. Altarelli and F. Feruglio, in *Neutrino Mass, Springer Tracts in Modern Physics* (G. Altarelli and K. Winter, eds.), hep-ph/0206077 (2002)

M. Battaglia et al., The CKM Matrix and the Unitarity Triangle, in the Workshop on CKM Unitarity Triangle (CERN 2002–2003), Geneva, Switzerland, 13–16 Feb 2002, hep-ph/0304132 (2002)

A. Buras, in *International School of Subnuclear Physics: 38th Course: Theory and Experiment Heading for New Physics*, Erice, 2000, hep-ph/0101336

T.E. Clark and S.T. Love, Mod. Phys. Lett. **A19**, 297 (2004)

G. Cvetic, C.O. Dib, C.S. Kim and J.D. Kim, Phys. Rev. D **71**, 113013 (2005)

K. Eguchi et al., Phys. Rev. Lett. **90**, 021802 (2003), hep-ex/0212021

G.L. Fogli, E. Lisi, A. Marrone, D. Montanino and A. Palazzo, Nucl. Phys. Proc. Suppl. **111**, 106 (2002)

Y. Fukuda et al., Phys. Rev. Lett. **81**, 1562 (1988), hep-ex/9807003

M. Gell-Mann, P. Ramond and R. Slansky in *Supergravity* (P. van Nieuwenhuizen and D. Freedman eds.) North-Holland, Amsterdam, 1979

S. Glashow, in *Proceedings of Neutrino 79, International Conference on Neutrinos, Weak Interactions and Cosmology*, Bergen, Norway, 1979, ed. by A. Haatuft and C. Jarlskog (Fysik Institutt, Bergen, 1980);

S.L. Glashow, in *Proceedings of the 1979 Cargese Institute on Quarks and Leptons* (M. Levy et al., eds.) Plenum Press, New York 1980, p687

M.C. Gonzalez-Garcia and Y. Nir, Rev. Mod. Phys. **75**, 345 (2003)

A. Halprin, Phys. Rev. Lett. **48**, 1313 (1982)

A. Halprin and A. Masiero, Phys. Rev. D **48**, R2987 (1993)

P. Herczeg and R.N. Mohapatra, Phys. Rev. Lett. **69**, 2475 (1992)

T. Huber et al., Phys. Rev. D **41**, 2709 (1990)

T.K. Kuo and S.T. Love, Phys. Rev. Lett. **45**, 93 (1980)

B. Liu, Gauge Invariance of the Muonium–Antimuonium Oscillation Time Scale and Limits on Right-Handed Neutrino Masses (2008)

R.E. Marshak and R.N. Mohapatra, Phys. Rev. Lett. **44**, 1316 (1980);

B. Matthias et al., Phys. Rev. Lett. **66**, 2716 (1991)

P. Minkowski, Phys. Lett. **B67**, 421 (1977)

R.N. Mohapatra and G. Senjanovic. Phys. Rev. Lett. **44**, 912 (1980)

B. Pontecorvo, Sov. Phys. JETP 6, 429 (1957)

L. Willmann et al., Phys. Rev. Lett. **82**, 49 (1999)

T. Yanagida, in *Proceedings of the Workshop on the Unified Theories and Baryon Number in the Universe*, Tsukuba, Japan, 1979, (O. Sawada and A. Sugamoto eds.), p95

the measurements of the [illegible] and [illegible] electron anomalous magnetic moments, respectively. Comparison shows that the bound from muon anomalous magnetic moment is more stringent than ones from [illegible] and electron anomalous magnetic moments. Finally, Chap. 6 presents the conclusions.

References

[illegible]
[illegible] Phys. Rev. Lett. 87, [illegible] (2001) [illegible]
[illegible] Phys. Rev. [illegible] (2006) [illegible]
M. [illegible] et al., [illegible] (2007)
G. [illegible] Report on [illegible]
M. [illegible]
A. [illegible]
[illegible] Phys. Rev. [illegible]
K. [illegible] Phys. Rev. [illegible]
[illegible] Phys. Rev. Lett. [illegible] (1988) [illegible]
[illegible] North-Holland, Amsterdam [illegible]
[illegible]
[illegible] Phys. Rev. Lett. [illegible]
[illegible] Phys. Rev. Lett. 45, [illegible] (1980)
[illegible]
[illegible] Phys. Rev. Lett. 44, [illegible]
[illegible] Phys. Rev. Lett. 66, [illegible] (1991)
[illegible] Phys. Lett. B67, [illegible] (1977)
[illegible] Phys. Rev. Lett. 42, [illegible] (1979)
[illegible]
[illegible] Phys. Rev. Lett. 22, [illegible]
[illegible]

Chapter 2
Muonium–Antimuonium Oscillation in a Modified Standard Model

2.1 Neutrino Masses and Mixings

We modify the Standard Model only by the inclusion of singlet right-handed neutrinos and allowing for general renormalizable interactions. This implies non-zero neutrino masses and mixing matrix elements. The leptonic Yukawa interactions with the Higgs scalar doublet in the modified Standard Model take the form

$$\mathcal{L}_{\text{int}}^{\phi} = -\frac{g}{\sqrt{2}M_W}\phi^{-}\left(\sum_{a,b=1}^{3}\overline{\ell_{Ra}^{(0)}}m_{ab}^{\ell}\nu_{Lb}^{(0)} - \overline{\ell_{La}^{(0)}}m_{ab}^{D\,\dagger}\nu_{Rb}^{(0)}\right) + H.C. \tag{2.1}$$

Here, $\ell_{La}^{(0)}$ and $\nu_{La}^{(0)}$ are, respectively, the charged lepton and its associated neutrino partner of the $SU(2)_L$ doublet, while $\nu_{Ra}^{(0)}$ is the right-handed neutrino singlet. The superscript zero indicates weak interaction eigenstates so that the leptonic charged current interaction is

$$\mathcal{L}_{\text{int}}^{W} = -\frac{g}{\sqrt{2}}W^{-\mu}\sum_{a=1}^{3}\overline{\ell_{La}^{(0)}}\gamma_{\mu}\nu_{La}^{(0)} + H.C. \tag{2.2}$$

After spontaneous symmetry breaking, the mass term for the charged leptons takes the form

$$\mathcal{L}_{\text{mass}}^{\ell} = -\sum_{a,b=1}^{3}\left[\overline{\ell_{Ra}^{(0)}}m_{ab}^{\ell}\ell_{Lb}^{(0)} + \overline{\ell_{La}^{(0)}}m_{ba}^{\ell *}\ell_{Rb}^{(0)}\right] \tag{2.3}$$

where m^{ℓ} is a 3×3 mass matrix. To diagonalize this matrix, one performs the biunitary transformation

$$m^{\ell} = A^{R}m_{\text{diag}}^{\ell}(A^{L})^{\dagger} \tag{2.4}$$

B. Liu, *Muonium–Antimuonium Oscillations in an Extended Minimal Supersymmetric Standard Model*, Springer Theses,
DOI: 10.1007/978-1-4419-8330-5_2, © Springer Science+Business Media, LLC 2011

where A^R and A^L are 3×3 unitary matrices and $m^{\ell}_{\rm diag}$ is a diagonal matrix, the entries of which are the charged lepton masses. To implement this basis change, the charged lepton fields participating in the weak interaction are rewritten in terms of the mass diagonal fields as

$$\ell^{(0)}_{La} = \sum_{b=1}^{3} A^L_{ab}\ell_{Lb}, \quad \ell^{(0)}_{Ra} = \sum_{b=1}^{3} A^R_{ab}\ell_{Rb} \tag{2.5}$$

On doing so, the mass term reads

$$\mathcal{L}^{\ell}_{\rm mass} = -\sum_{a=1}^{3} m_{\ell a}\left[\overline{\ell_{Ra}}\ell_{Lb} + \overline{\ell_{La}}\ell_{Rb}\right] \tag{2.6}$$

A general neutrino mass term resulting from renormalizable interactions takes the form

$$\mathcal{L}^{\nu}_{\rm mass} = -\frac{1}{2}\overline{\left((\nu^{(0)}_L)^c \nu^{(0)}_R\right)}\begin{pmatrix} 0 & (m^D)^T \\ m^D & m^R \end{pmatrix}\begin{pmatrix} \nu^{(0)}_L \\ (\nu^{(0)}_R)^c \end{pmatrix} + H.C. \tag{2.7}$$

Note that the upper left 3×3 block in the neutrino mass matrix is set to zero. This block matrix involves only left-handed neutrinos and in the (modified) Standard Model its generation requires a nonrenormalizable mass dimension-five operator. Consequently, such a term will be ignored.

For three generations of neutrinos, the six mass eigenvalues, $m_{\nu A}$, are obtained from the diagonalization of the 6×6 matrix

$$M^{\nu} = \begin{pmatrix} 0 & (m^D)^T \\ m^D & m^R \end{pmatrix} \tag{2.8}$$

Since M^{ν} is symmetric, it can be diagonalized by a single unitary 6×6 matrix, U, as

$$M^{\nu}_{\rm diag} = U^T M^{\nu} U. \tag{2.9}$$

This diagonalization is implemented via the basis change on the original neutrino fields organized as the six-dimensional column vector

$$N^{(0)}_L = \begin{pmatrix} \nu^{(0)}_L \\ (\nu^{(0)}_R)^c \end{pmatrix}, \quad N^{(0)}_R = \begin{pmatrix} (\nu^{(0)}_L)^c \\ \nu^{(0)}_R \end{pmatrix} \tag{2.10}$$

to the new neutrino fields defined as

$$N^{(0)}_L = U N_L, \quad N^{(0)}_R = U^* N_R \tag{2.11}$$

where

$$N_L = \begin{pmatrix} \nu_L \\ (\nu_R)^c \end{pmatrix}, \quad N_R = \begin{pmatrix} (\nu_L)^c \\ \nu_R \end{pmatrix}. \tag{2.12}$$

The neutrino mass term then takes the form

$$\mathcal{L}^{\nu}_{\text{mass}} = -\frac{1}{2}\sum_{A=1}^{6} m_{\nu A}\left[\nu_A^T C \nu_A + \overline{\nu_A} C \overline{\nu_A^T}\right] = -\sum_{A=1}^{6} m_{\nu A}\overline{\nu_A}\nu_A, \tag{2.13}$$

where $m_{\nu A}$ are the Majorana neutrino masses.

Since a nonzero Majorana mass matrix m^R does not require $SU(2)_L \times U(1)$ symmetry breaking, it is naturally characterized by a much larger scale, M_R, than the elements of the matrix m^D the nontrivial values of which require $SU(2)_L \times U(1)$ symmetry breaking and are thus expected to be somewhere in the order of the charged lepton mass to the W mass. Thus, one can take the elements of m^D, characterized by a scale m_D, to be much less than M_R, the scale of the elements of m^R. One then finds on diagonalization of the 6×6 neutrino mass matrix that three of the eigenvalues are crudely given by

$$m_{\nu a} \sim \frac{m_D^2}{M_R} \ll m_D, \quad a = 1,2,3, \tag{2.14}$$

while the other three eigenvalues are roughly

$$m_{\nu i} \sim M_R, \quad i = 4,5,6. \tag{2.15}$$

This constitutes the so-called see-saw mechanism (Minkowski 1977; Yanagida 1979; Gell-Mann et al. 1979; Glashow 1979; Mohapatra and Senjanovic 1980) and provides a natural explanation of the smallness of the three light neutrino masses. Moreover, the elements of the mixing matrix are characterized by an M_R mass dependence

$$\begin{aligned} &U_{ab} \sim \mathcal{O}(1), \quad a,b = 1,2,3 \\ &U_{ij} \sim \mathcal{O}(1), \quad i,j = 4,5,6 \\ &U_{ia} \sim U_{ai} \sim \mathcal{O}\left(\frac{m_D}{M_R}\right), \quad a = 1,2,3,\ i = 4,5,6. \end{aligned} \tag{2.16}$$

Since the charged lepton mixing matrix is independent of M_R, one finds that elements of the mixing matrix appearing in the charged current have the M_R mass dependence

$$\begin{aligned} &V_{ab} \sim \mathcal{O}(1), \quad a,b = 1,2,3 \\ &V_{ai} \sim \mathcal{O}\left(\frac{m_D}{M_R}\right), \quad a = 1,2,3,\ i = 4,5,6 \end{aligned} \tag{2.17}$$

Inserting the transformations (2.5) and (2.11) in the interaction terms (2.1) and (2.2), and taking into account the mass matrix transformations (2.4) and (2.9), one obtains the explicit interactions of charged bosons with the leptons in their mass diagonal basis as

$$\mathcal{L}_{\text{int}}^{W} = -\frac{g}{\sqrt{2}} W^{-\mu} \sum_{a=1}^{3} \sum_{A=1}^{6} \bar{\ell}_{La} \gamma_{\mu} V_{aA} \nu_A - \frac{g}{\sqrt{2}} W^{+\mu} \sum_{a=1}^{3} \sum_{A=1}^{6} \bar{\nu}_A V_{aA}^{*} \gamma_{\mu} \ell_{La} \tag{2.18}$$

$$\begin{aligned} \mathcal{L}_{\text{int}}^{\phi} = & -\frac{g}{\sqrt{2} M_W} \phi^{-} \sum_{a=1}^{3} \sum_{A=1}^{6} \bar{\ell}_a V_{aA} \left(m_{la} \frac{1-\gamma_5}{2} - m_{\nu A} \frac{1+\gamma_5}{2} \right) \nu_A \\ & - \frac{g}{\sqrt{2} M_W} \phi^{+} \sum_{a=1}^{3} \sum_{A=1}^{6} \bar{\nu}_A V_{aA}^{*} \left(m_{la} \frac{1+\gamma_5}{2} - m_{\nu A} \frac{1-\gamma_5}{2} \right) \ell_a \end{aligned} \tag{2.19}$$

where

$$V_{aA} = \sum_{b=1}^{3} (A_L^{-1})_{ab} U_{bA} \tag{2.20}$$

Note that the mixing matrix V_{aA} satisfies the identities (Ilakovac and Pilaftsis 1995):

$$\sum_{A=1}^{6} V_{aA} V_{bA}^{*} = \delta_{ab} \tag{2.21}$$

$$\sum_{A=1}^{6} V_{aA} V_{bA} m_{\nu A} = 0 \tag{2.22}$$

Identity (2.21) stems from the unitarity of matrices A_L and U, while identity (2.22) is a consequence of the particular form of the neutrino mass matrix. In particular, it requires the Majorana mass term of the left-handed neutrinos be set to zero. A detailed proof of this latter identity is provided in Appendix A.

The propagators in R_ξ gauge are (Cheng and Li 1988):

$$\begin{aligned} &\text{Fermions}: \frac{i(\gamma p + m)}{p^2 - m^2 + i\epsilon} \\ &\text{W-bosons } W^{\pm}: \frac{-i}{p^2 - M_W^2 + i\epsilon} \left[g_{\mu\nu} + \frac{(\xi - 1) p_\mu p_\nu}{p^2 - \xi M_W^2} \right] \\ &\text{Goldstone bosons } \phi^{\pm}: \frac{i}{p^2 - \xi M_W^2} \end{aligned} \tag{2.23}$$

2.2 The Gauge Invariant T-Matrix Elements of Muonium–Antimuonium Oscillation

The lowest order Feynman diagrams accounting for muonium and antimuonium mixing are displayed in Fig. 2.1. We shall consistently employ the R_ξ gauge. The gauge invariance of the T-matrix element will be demonstrated by establishing its ξ independence. In Fig. 2.1, there are two neutrinos in the intermediate state for each graph, while every wavy line represents either a W boson or an R_ξ gauge charged erstwhile Nambu-Goldstone boson.

Note that in the unitary gauge ($\xi \to \infty$), the W boson propagator takes the form $\frac{-i}{p^2-M_W^2+i\epsilon}[g_{\mu\nu} - \frac{p_\mu p_\nu}{M_W^2}]$. A theory with such a propagator has very bad power counting convergence properties. As it turns out, the unitary gauge power counting divergent pieces in the W vector box diagrams vanish, as they must, after application of properties (2.21) and (2.22). Hence, when we calculate the T-matrix elements in R_ξ gauge, we will also apply properties (2.21) and (2.22) to establish the cancelation of the various terms in this case.

As it turns out, graph (a) gives the same contribution as (b), as do graphs (c) and (d). Hence, we need only discuss the gauge invariant T-matrix elements of the

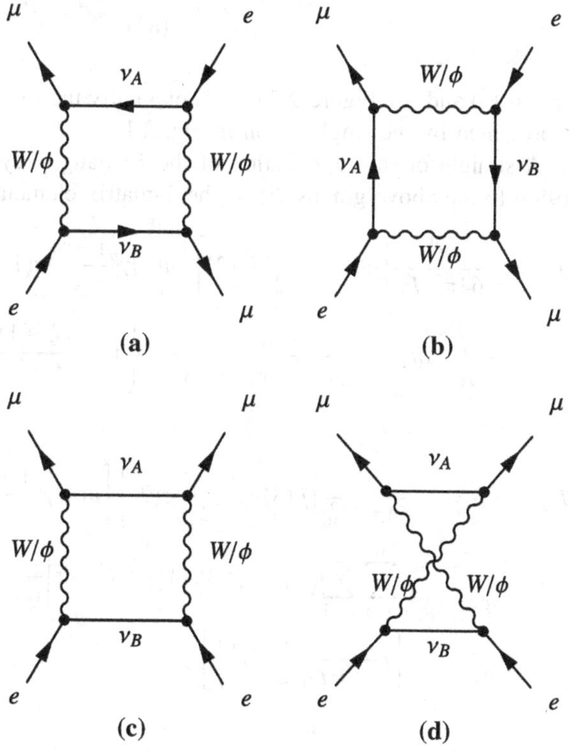

Fig. 2.1 Feynman graphs contributing to the muonium–antimuonium mixing. Each wavy line is either a W boson or an R_ξ gauge charged Nambu–Goldstone boson

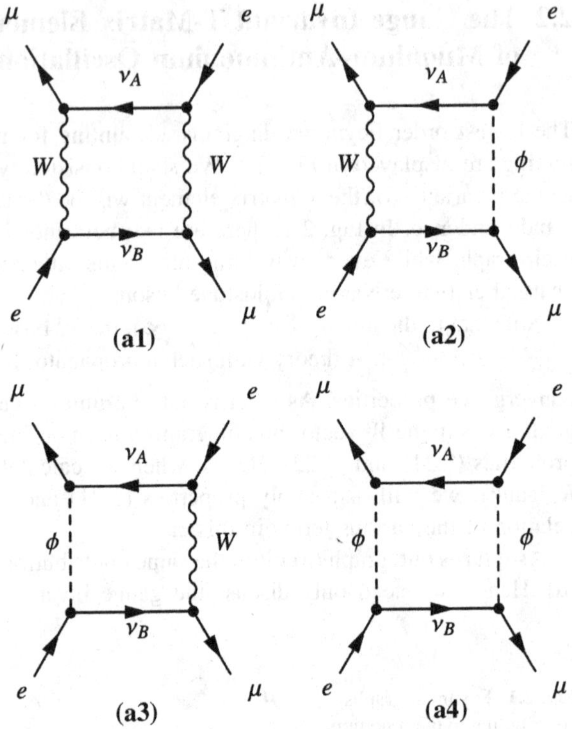

Fig. 2.2 Feynman graphs of type (**a**) in R_ξ gauge, which are presented as one single graph (**a**) in Fig. 2.1

graphs (a) and (c). Figure 2.2 details explicitly the four separate graphs, which are represented by the single graph in Fig. 2.1.

A straightforward application of the R_ξ gauge Feynman rules (Cheng and Li 1988) to the above graphs yields the T-matrix elements

$$T_{a1} = -\frac{g^4}{64\pi^2 M_W^2}\left[\bar{\mu}(3)\gamma_\mu\frac{1-\gamma_5}{2}e(2)\right]\left[\bar{\mu}(4)\gamma^\mu\frac{1-\gamma_5}{2}e(1)\right]\sum_{A=1}^{6}\sum_{B=1}^{6}(V_{\mu A}V_{eA}^*)(V_{\mu B}V_{eB}^*)$$
$$\cdot\int_0^\infty dt\left[\frac{x_A x_B}{(t+x_A)(t+x_B)(t+1)^2}\cdot\left\{1+\frac{2(\xi-1)}{t+\xi}\cdot t+\frac{(\xi-1)^2}{4(t+\xi)^2}\cdot t^2\right\}\right] \tag{2.24}$$

$$T_{a2} = T_{a3} = -\frac{g^4}{64\pi^2 M_W^2}\left[\bar{\mu}(3)\gamma_\mu\frac{1-\gamma_5}{2}e(2)\right]\left[\bar{\mu}(4)\gamma^\mu\frac{1-\gamma_5}{2}e(1)\right]$$
$$\cdot\sum_{A=1}^{6}\sum_{B=1}^{6}(V_{\mu A}V_{eA}^*)(V_{\mu B}V_{eB}^*)\int_0^\infty dt\left[\frac{x_A x_B}{(t+x_A)(t+x_B)(t+1)(t+\xi)}\right.$$
$$\left.\cdot\left\{t+\frac{\xi-1}{4(t+\xi)}\cdot t^2\right\}\right] \tag{2.25}$$

$$T_{a4} = -\frac{g^4}{64\pi^2 M_W^2}\left[\bar{\mu}(3)\gamma_\mu \frac{1-\gamma_5}{2} e(2)\right]\left[\bar{\mu}(4)\gamma^\mu \frac{1-\gamma_5}{2} e(1)\right]\sum_{A=1}^{6}\sum_{B=1}^{6}(V_{\mu A}V_{eA}^*)(V_{\mu B}V_{eB}^*)$$
$$\cdot \int_0^\infty dt \left[\frac{x_A x_B}{(t+x_A)(t+x_B)(t+\xi)^2}\cdot\frac{t^2}{4}\right] \tag{2.26}$$

where $\bar{\mu}(3) = \bar{\mu}(p_3, s_3)$, $\bar{\mu}(4) = \bar{\mu}(p_4, s_4)$, $e(1) = e(p_1, s_1)$ and $e(2) = e(p_2, s_2)$ are the spinors of the muons and electrons and $x_A = \frac{m_{\nu A}^2}{M_W^2}$, $A = 1, \ldots, 6$. Note that in obtaining these results, we already applied properties (2.21) and (2.22) to eliminate various self-canceling terms. As such, the integrals in (2.24)–(2.26) are finite even in the $\xi \to \infty$ limit. See Appendix B for detailed calculations of T-matrix elements.

In order to discuss the ξ dependence in a manifest way, we rewrite these T-matrix elements as

$$\begin{aligned}
T_{a1} &= \sum_{A=1}^{6}\sum_{B=1}^{6}\int_0^\infty dt\mathcal{A}(x_A, x_B, t)\cdot\left(h(t)\cdot\frac{1}{(t+\xi)^2} - 2g(t)\cdot\frac{1}{t+\xi} + f(t)\right)\\
T_{a2} &= \sum_{A=1}^{6}\sum_{B=1}^{6}\int_0^\infty dt\mathcal{A}(x_A, x_B, t)\cdot\left(-h(t)\cdot\frac{1}{(t+\xi)^2} + g(t)\cdot\frac{1}{t+\xi}\right)\\
T_{a3} &= \sum_{A=1}^{6}\sum_{B=1}^{6}\int_0^\infty dt\mathcal{A}(x_A, x_B, t)\cdot\left(-h(t)\cdot\frac{1}{(t+\xi)^2} + g(t)\cdot\frac{1}{t+\xi}\right)\\
T_{a4} &= \sum_{A=1}^{6}\sum_{B=1}^{6}\int_0^\infty dt\mathcal{A}(x_A, x_B, t)\cdot\left(h(t)\cdot\frac{1}{(t+\xi)^2}\right)
\end{aligned} \tag{2.27}$$

where

$$\mathcal{A}(x_A, x_B, t) = -\frac{g^4}{64\pi^2 M_W^2}\left[\bar{\mu}(3)\gamma_\mu \frac{1-\gamma_5}{2} e(2)\right]\left[\bar{\mu}(4)\gamma^\mu \frac{1-\gamma_5}{2} e(1)\right](V_{\mu A}V_{eA}^*)(V_{\mu B}V_{eB}^*)$$
$$\cdot\frac{x_A x_B}{(t+x_A)(t+x_B)} \tag{2.28}$$

with

$$h(t) = \frac{t^2}{4}, \quad g(t) = \frac{t+\frac{t^2}{4}}{t+1} \text{ and } f(t) = \frac{1+2t+\frac{t^2}{4}}{(t+1)^2} \tag{2.29}$$

Note that the $\frac{1}{(t+\xi)^2}$ terms from the second and third graphs totally cancel against the ones from the first and fourth graphs, while the $\frac{1}{t+\xi}$ terms from the second and third graphs exactly cancel the one from the first graph. All ξ dependent contributions thus vanish and the only remaining piece is the term containing $f(t)$

from the first graph, which is ξ independent. Hence, we have the gauge invariant T-matrix element for graphs of type (a)

$$\begin{aligned}T_a =& -\frac{g^4}{64\pi^2 M_W^2}\left[\bar{\mu}(3)\gamma_\mu \frac{1-\gamma_5}{2} e(2)\right]\left[\bar{\mu}(4)\gamma^\mu \frac{1-\gamma_5}{2} e(1)\right] \\ & \cdot \sum_{A=1}^{6}\sum_{B=1}^{6}(V_{\mu A}V_{eA}^*)(V_{\mu B}V_{eB}^*)x_A x_B \cdot \int_0^\infty dt \frac{1+2t+\frac{t^2}{4}}{(t+x_A)(t+x_B)(t+1)^2} \\ =& -\frac{G_F^2 M_W^2}{8\pi^2}[\bar{\mu}(3)\gamma_\mu(1-\gamma_5)e(2)][\bar{\mu}(4)\gamma^\mu(1-\gamma_5)e(1)] \\ & \cdot \left[\sum_{A=1}^{6}(V_{\mu A}V_{eA}^*)^2 S(x_A) + \sum_{A,B=1;A\neq B}^{6}(V_{\mu A}V_{eA}^*)(V_{\mu B}V_{eB}^*)T(x_A,x_B)\right]\end{aligned} \tag{2.30}$$

Here, we have introduced the Fermi scale

$$\frac{G_F}{\sqrt{2}} = \frac{g^2}{8M_W^2} \tag{2.31}$$

along with the Inami–Lim (Inami and Lim 1981) function

$$S(x_A) = \frac{x^3 - 11x^2 + 4x}{4(1-x)^2} - \frac{3x^3}{2(1-x)^3}\ln(x) \tag{2.32}$$

We have also defined

$$T(x_A, x_B) = x_A x_B \left(\frac{J(x_A) - J(x_B)}{x_A - x_B}\right) = T(x_B, x_A) \tag{2.33}$$

with

$$J(x) = \frac{x^2 - 8x + 4}{4(1-x)^2}\ln(x) - \frac{3}{4}\frac{1}{(1-x)} \tag{2.34}$$

In a similar manner, the graph (c) in Fig. 2.1 represents the four graphs in Fig. 2.3.

The T-matrix elements of these four graphs are

$$\begin{aligned} T_{c1} &= \sum_{A=1}^{6}\sum_{B=1}^{6}\int_0^\infty dt\mathcal{B}(x_A, x_B, t) \cdot \left(t \cdot \frac{1}{(t+\xi)^2} - 2 \cdot \frac{1}{t+\xi} + \tilde{f}(x_A, x_B, t)\right) \\ T_{c2} &= \sum_{A=1}^{6}\sum_{B=1}^{6}\int_0^\infty dt\mathcal{B}(x_A, x_B, t) \cdot \left(-t \cdot \frac{1}{(t+\xi)^2} + \frac{1}{t+\xi}\right) \\ T_{c3} &= \sum_{A=1}^{6}\sum_{B=1}^{6}\int_0^\infty dt\mathcal{B}(x_A, x_B, t) \cdot \left(-t \cdot \frac{1}{(t+\xi)^2} + \frac{1}{t+\xi}\right) \\ T_{c4} &= \sum_{A=1}^{6}\sum_{B=1}^{6}\int_0^\infty dt\mathcal{B}(x_A, x_B, t) \cdot \left(t \cdot \frac{1}{(t+\xi)^2}\right) \end{aligned} \tag{2.35}$$

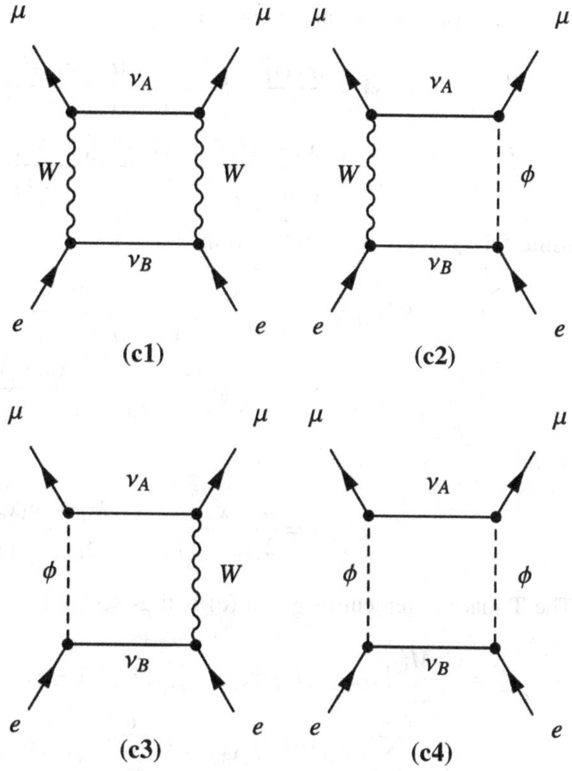

Fig. 2.3 Feynman graphs of type (**c**) in R_ξ gauge, which are presented as one single graph (**a**) in Fig. 2.1

where

$$\mathcal{B}(x_A, x_B, t) = \frac{g^4}{64\pi^2 M_W^4}\left[\bar{\mu}(3)\gamma^\mu \frac{1-\gamma_5}{2} e(2)\right]\left[\bar{\mu}(4)\gamma_\mu \frac{1-\gamma_5}{2} e(1)\right](V_{\mu A})^2 (V_{eB}^*)^2 \cdot \frac{m_A \cdot m_B}{2} \cdot \frac{x_A x_B}{(t+x_A)(t+x_B)} \tag{2.36}$$

and

$$\tilde{f}(x_A, x_B, t) = \frac{\frac{4t}{x_A \cdot x_B} + t + 2}{(t+1)^2} \tag{2.37}$$

In a similar fashion to the case for the graphs of type (a), all the ξ dependent terms again cancel against each other leaving only the ξ independent $\tilde{f}(x_A, x_B, t)$ term. Thus the type (c) T-matrix element is gauge invariant and is given by

$$T_c = \frac{G_F^2 M_W^2}{8\pi^2}[\bar{\mu}(3)\gamma^\mu(1-\gamma_5)e(2)][\bar{\mu}(4)\gamma_\mu(1-\gamma_5)e(1)]\sum_{A=1}^{6}(V_{\mu A})^2 \sum_{B=1}^{6}(V_{eB}^*)^2 \frac{\sqrt{x_A x_B}}{2} \cdot \int_0^\infty dt \left\{\frac{4t + x_A x_B(t+2)}{(t+x_A)(t+x_B)(t+1)^2}\right\} \tag{2.38}$$

If $x_A = x_B$, the relevant integral is

$$I(x_A) = \frac{\sqrt{x_A x_B}}{2} \cdot \int_0^\infty dt \frac{4t + x_A x_B(t+2)}{(t+x_A)(t+x_B)(t+1)^2} = \frac{(x_A-4)x_A}{(x_A-1)^2} + \frac{(x_A^3 - 3x_A^2 + 4x_A + 4)x_A}{2(x_A-1)^3} \ln x_A \tag{2.39}$$

while for $x_A \neq x_B$, it takes the form

$$K(x_A, x_B) = \frac{\sqrt{x_A x_B}}{2} \cdot \int_0^\infty dt \frac{4t + x_A x_B(t+2)}{(t+x_A)(t+x_B)(t+1)^2} = \sqrt{x_A x_B} \frac{L(x_A, x_B) - L(x_B, x_A)}{x_A - x_B} \tag{2.40}$$

with

$$L(x_A, x_B) = \frac{4 - x_A x_B}{2(x_A - 1)} + \frac{x_A(2x_B - x_A x_B - 4)}{2(x_A-1)^2} \ln x_A \tag{2.41}$$

The T-matrix element of graph (c) is thus secured as

$$T_c = \frac{G_F^2 M_W^2}{8\pi^2} [\bar{\mu}(3)\gamma^\mu(1-\gamma_5)e(2)] [\bar{\mu}(4)\gamma_\mu(1-\gamma_5)e(1)] \cdot \left[\sum_{A=1}^{6} (V_{\mu A} V_{eA}^*)^2 I(x_A) + \sum_{A,B=1;A\neq B}^{6} (V_{\mu A})^2 (V_{eB}^*)^2 K(x_A, x_B) \right] \tag{2.42}$$

2.3 The Effective Lagrangian

Combining the various contributions, the T-matrix element can be reproduced using the gauge invariant effective Lagrangian given by:

$$\mathcal{L}_{\text{eff}} = \frac{G_{\bar{M}M}}{\sqrt{2}} [\bar{\mu}\gamma^\mu(1-\gamma_5)e] [\bar{\mu}\gamma_\mu(1-\gamma_5)e] \tag{2.43}$$

where

$$\frac{G_{\bar{M}M}}{\sqrt{2}} = -\frac{G_F^2 M_W^2}{16\pi^2} \left[\sum_{A=1}^{6} (V_{\mu A} V_{eA}^*)^2 S(x_A) + \sum_{A,B=1;A\neq B}^{6} (V_{\mu A} V_{eA}^*)(V_{\mu B} V_{eB}^*) T(x_A, x_B) - \sum_{A=1}^{6} (V_{\mu A} V_{eA}^*)^2 I(x_A) - \sum_{A,B=1;A\neq B}^{6} (V_{\mu A})^2 (V_{eB}^*)^2 K(x_A, x_B) \right]$$

$$= -\frac{G_F^2 M_W^2}{16\pi^2}\left[\sum_{A=1}^{6}(V_{\mu A}V_{eA}^*)^2(S(x_A) - I(x_A))\right.$$
$$\left. + \sum_{A,B=1;A\neq B}^{6}\Big((V_{\mu A}V_{eA}^*)(V_{\mu B}V_{eB}^*)T(x_A,x_B) - (V_{\mu A})^2(V_{eB}^*)^2K(x_A,x_B)\Big)\right] \tag{2.44}$$

2.4 Limit on M_R

Muonium (antimuonium) is a nonrelativistic Coulombic bound state of an electron and an anti-muon (positron and muon). The nontrivial mixing between the muonium ($|M\rangle$) and antimuonium ($|\bar{M}\rangle$) states is encapsulated in the effective Lagrangian of Eq. (2.43) and leads to the mass diagonal states given by the linear combinations (see Appendix C for detailed derivation)

$$|M_\pm\rangle = \frac{1}{\sqrt{2(1+|\varepsilon|^2)}}[(1+\varepsilon)|M\rangle \pm (1-\varepsilon)|\bar{M}\rangle] \tag{2.45}$$

where

$$\varepsilon = \frac{\sqrt{\mathcal{M}_{M\bar{M}}} - \sqrt{\mathcal{M}_{\bar{M}M}}}{\sqrt{\mathcal{M}_{M\bar{M}}} + \sqrt{\mathcal{M}_{\bar{M}M}}} \tag{2.46}$$

$$\mathcal{M}_{M\bar{M}} = \frac{\langle M| - \int d^3r\mathcal{L}_{\text{eff}}|\bar{M}\rangle}{\sqrt{\langle M|M\rangle\langle\bar{M}|\bar{M}\rangle}}, \quad \mathcal{M}_{\bar{M}M} = \frac{\langle\bar{M}| - \int d^3r\mathcal{L}_{\text{eff}}|M\rangle}{\sqrt{\langle M|M\rangle\langle\bar{M}|\bar{M}\rangle}} \tag{2.47}$$

Since the neutrino sector is expected to be CP violating, these will be independent, complex matrix elements. If the neutrino sector conserves CP, with $|M\rangle$ and $|\bar{M}\rangle$ CP conjugate states, then $\mathcal{M}_{M\bar{M}} = \mathcal{M}_{\bar{M}M}$ and $\epsilon = 0$. In general, the magnitude of the mass splitting between the two mass eigenstates is

$$|\Delta M| = 2\left|\text{Re}\sqrt{\mathcal{M}_{M\bar{M}}\mathcal{M}_{\bar{M}M}}\right| \tag{2.48}$$

Since muonium and antimuonium are linear combinations of the mass diagonal states, an initially prepared muonium or antimuonium state will undergo oscillations into one another as a function of time. The muonium–antimuonium oscillation timescale, $\tau_{\bar{M}M}$, is given by

$$\frac{1}{\tau_{\bar{M}M}} = |\Delta M|. \tag{2.49}$$

We would like to evaluate $|\Delta M|$ in the nonrelativistic limit. A nonrelativistic reduction of the effective Lagrangian of Eq. (2.43) produces a local, complex effective potential

$$V_{\text{eff}}(\mathbf{r}) = 8\frac{G_{\bar{M}M}}{\sqrt{2}}\delta^3(\mathbf{r}) \tag{2.50}$$

Taking the muonium (antimuonium) to be in their respective Coulombic ground states, $\phi_{100}(\mathbf{r}) = \frac{1}{\sqrt{\pi a_{\bar{M}M}^3}}e^{-r/a_{\bar{M}M}}$, where $a_{\bar{M}M} = \frac{1}{m_{\text{red}}\alpha}$ is the muonium Bohr radius with $m_{\text{red}} = \frac{m_e m_\mu}{m_e+m_\mu} \simeq m_e$ the reduced mass of muonium, it follows that

$$\begin{aligned}\frac{1}{\tau_{\bar{M}M}} &\simeq 2^3 r\phi_{100}^*(\mathbf{r})|\text{Re}V_{\text{eff}}(\mathbf{r})|\phi(\mathbf{r})_{100} \\ &= 16\frac{|\text{Re}G_{\bar{M}M}|}{\sqrt{2}}|\phi_{100}(0)|^2 = \frac{16|\text{Re}G_{\bar{M}M}|}{\pi}\frac{1}{\sqrt{2}}\frac{1}{a_{\bar{M}M}^3}\end{aligned} \tag{2.51}$$

Thus, we secure an oscillation timescale

$$\frac{1}{\tau_{\bar{M}M}} \simeq \frac{16|\text{Re}G_{\bar{M}M}|}{\pi\sqrt{2}}m_e^3\alpha^3 \tag{2.52}$$

The present experimental limit (Willmann et al. 1999) on the non-observation of muonium–antimuonium oscillation translates into the bound $|\text{Re}G_{\bar{M}M}| \leq 3.0 \times 10^{-3}G_F$ where $G_F \simeq 1.16 \times 10^{-5}\,\text{GeV}^{-2}$ is the Fermi scale. This limit can then be used to construct a crude lower bound on M_R. For the case when the neutrino masses arise from a see-saw mechanism and taking m_D to be of order M_W, the M_R dependence of $G_{\bar{M}M}$ is obtained from Eq. (2.44) as:

$$\begin{aligned}&\textit{Case}\,1: \quad |\text{Re}G_{\bar{M}M}| \sim \frac{G_F^2 M_W^4}{M_R^2}\ln\frac{M_R}{M_W}, \quad A = 1,2,3, \quad B = 1,2,3 \\ &\textit{Case}\,2: \quad |\text{Re}G_{\bar{M}M}| \sim \frac{G_F^2 M_W^4}{M_R^2}\ln\frac{M_R}{M_W}, \quad A = 4,5,6, \quad B = 4,5,6 \\ &\textit{Case}\,3: \quad |\text{Re}G_{\bar{M}M}| \sim \frac{G_F^2 M_W^6}{M_R^4}\ln\frac{M_R}{M_W}, \quad A = 1,2,3, \quad B = 4,5,6\end{aligned} \tag{2.53}$$

Cases 1 and 2 give the same order M_R dependence, while case 3 is suppressed by an additional factor of $\frac{M_W^2}{M_R^2}$. Hence, the term $\frac{G_F^2 M_W^4}{M_R^2}\ln\frac{M_R}{M_W}$ gives the dominant contribution. We then roughly calculate a bound of M_R as

$$\frac{G_F^2 M_W^4}{M_R^2}\ln\frac{M_R}{M_W} \leq 3.0 \times 10^{-3}G_F \tag{2.54}$$

which has also been obtained in reference (Cvetic et al. 2005). Using $M_W \simeq 80.4$ GeV and $G_F = 1.166 \times 10^{-5}\,\text{GeV}^{-2}$, we finally secure

$$M_R \geq 6 \times 10^2\,\text{GeV} \tag{2.55}$$

Note that this is just a rough estimate, since we are retaining only the dependence on M_R while neglecting all numerical dependence on the mixing angles and CP violating phases in V_{aA}.

References

T.-P. Cheng and L.-F. Li, Gauge theory of elementary particle physics, 1988, p. 509.

G. Cvetic, C.O. Dib, C.S. Kim and J.D. Kim, Phys. Rev. D **71**, 113013 (2005).

M. Gell-Mann, P. Ramond and R. Slansky in *Supergravity*, (P. van Nieuwenhuizen and D. Freedman eds) North-Holland, Amsterdam, 1979.

S.L. Glashow, in *Proceedings of the 1979 Cargese Institute on Quarks and Leptons* (M. Levy et. al. eds.) Plenum Press, New York 1980, p. 687.

A. Ilakovac and A. Pilaftisis, Nucl. Phys. **B437** (1995) 491.

T. Inami and C.S. Lim, Prog. Theor. Phys. **65**, 297 (1981).

P. Minkowski, Phys. Lett. **B67**, 421 (1977).

R.N. Mohapatra and G. Senjanovic. Phys. Rev. Lett. **44**, 912 (1980).

L. Willmann et al, Phys. Rev. Lett **82**, 49 (1999).

T. Yanagida, in *Proceedings of the Workshop on the Unified Theories and Baryon Number in the Universe*, Tsukuba, Japan, 1979, (O. Sawada and A. Sugamoto eds), p. 95.

Chapter 3
Supersymmetry and the Minimal Supersymmetric Standard Model

3.1 Why Supersymmetry?

1. One reason that physicists explore supersymmetry (SUSY) is because it offers an extension to the more familiar space–time symmetries of quantum field theory. These symmetries are grouped into the Poincaré group and internal symmetries. The Coleman–Mandula theorem (Coleman and Mandula 1967) showed that under certain assumptions, the symmetries of the S-matrix must be a direct product of the Poincaré group with a compact internal symmetry group or, if there is no mass gap, the conformal group with a compact internal symmetry group. In 1975, the Haag–Lopuszanski–Sohnius theorem (Haag et al. 1975) showed that considering symmetry generators that satisfy anticommutation relations allows for such nontrivial extensions of space–time symmetry. This extension of the Coleman–Mandula theorem prompted some physicists to study this wider class of theories.
2. One of the main motivations for SUSY comes from the quadratically divergent contributions to the Higgs mass squared. The quantum mechanical interactions of the Higgs boson causes a large renormalization of the Higgs mass and, unless there is an accidental cancelation or fine tuning, the natural size of the Higgs mass is the highest scale possible. However, taking all precision measurements together in a global fit, the current experiment infers that the Standard Model Higgs boson mass must be lighter than around 200 GeV (95% c.l.) (2006). This problem is known as the hierarchy problem (Weinberg 1976; Gildener 1976; Susskind 1979; t Hooft 1979). Consider a massive fermion loop correction to the propagator for the Higgs field, as shown in Fig. 3.1. If the Higgs field couples to a Dirac fermion f with a term in the Lagrangian $-\lambda_f H\bar{f}f$, the leading correction is then given by

B. Liu, *Muonium–Antimuonium Oscillations in an Extended Minimal Supersymmetric Standard Model*, Springer Theses,
DOI: 10.1007/978-1-4419-8330-5_3,

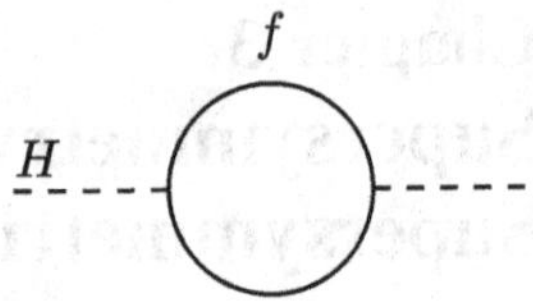

Fig. 3.1 Fermion loop contribution to the self-energy of the Higgs boson

$$\Delta m_H^2 = -\frac{|\lambda_f|^2}{8\pi^2}[\Lambda_{UV}^2 - 2m_f^2 \ln(\Lambda_{UV}/m_f) + \cdots] \tag{3.1}$$

Here, Λ_{UV} is an ultraviolet momentum cutoff used to regulate the loop integral. If the cutoff, Λ_{UV}, is replaced by the Planck mass, M_{planck}, the resulting correction would be some thirty degrees of magnitude larger than the experimental bound on the Higgs. Supersymmetry provides automatic cancelations between fermionic and bosonic Higgs interactions. For example, consider a scalar field S, which couples to Higgs with a Lagrangian term $-\lambda_S|H|^2|S|^2$. Then the Feynman graph in Fig. 3.2 gives a correction

$$\Delta m_H^2 = \frac{\lambda_S}{16\pi^2}[\Lambda_{UV}^2 - 2m_S^2 \ln(\Lambda_{UV}/m_s) + \cdots] \tag{3.2}$$

If each of the quarks and leptons of the Standard Model is accompanied by two complex scalars with $\lambda_S = |\lambda_f|^2$, then the Λ_{UV} contributions of Figs. 3.1, 3.2 will cancel. Supersymmetry relates the fermion and boson couplings in just this manner. Each Standard Model field gets a super partner with couplings to insure the cancelations.

3. Another motivation for supersymmetry existing at the weak scale is gauge coupling unification. The renormalization group evolution of the three gauge coupling constants of the Standard Model is sensitive to the particle content of the theory. These coupling constants do not quite meet together at a common energy scale if we run the renormalization group using the Standard Model. Supersymmetry actually allows for the unification of three other forces, strong, weak and electromagnetic as shown in Fig. 3.3 (Martin 1997).
4. The most general superpotential of MSSM contains terms where the baryon and the lepton numbers are violated. To present this, a new discrete symmetry, called R-parity can be introduced. It is a multiplicative quantum number where all the particles of the Standard Model have positive R-parity, while their superpartners have negative R-parity and the quantum number is given by

$$R = (-1)^{3(B-L)+2s} \tag{3.3}$$

for a particle with spin s and baryon and lepton number B and L. This would lead to a useful phenomenological result. The lightest s particle, called the **LSP**, must be absolutely stable. If it is electrically neutral, the **LSP** is an excellent candidate for dark matter (for a review, see Jungman et al. 1996).

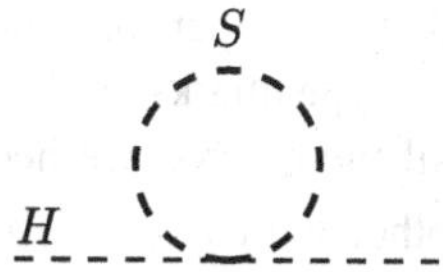

Fig. 3.2 Scalar loop contribution to the self-energy of the Higgs boson

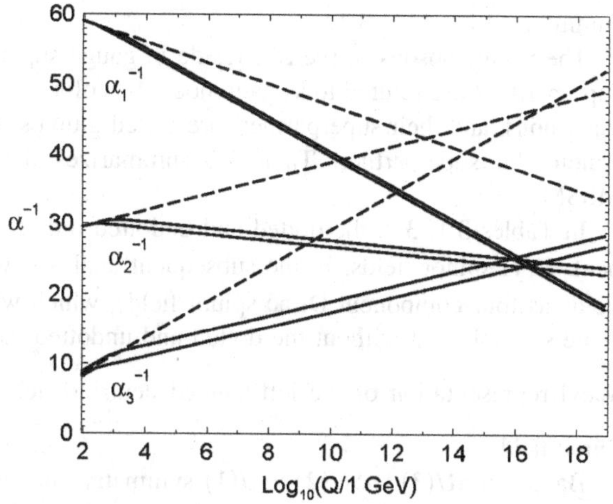

Fig. 3.3 Dashed lines denote the Standard Model couplings and solid lines denote the Minimal Supersymmetric Standard Model couplings

3.2 Minimal Supersymmetric Standard Model

The Minimal Supersymmetric Standard Model (MSSM) is the supersymmetric extension of the Standard Model with a minimal particle content (Martin 1997). For each particle, there is a superpartner with the same internal quantum numbers, but with spin that differs by half a unit.

All of the SM fermions have the property that the left-handed and right-handed components transform differently under the gauge groups. Only chiral multiplets can contain fermions in which left-handed components transform differently from their right-handed partners under a gauge group. Therefore in the MSSM, each of the fundamental particles must be in either a gauge or a chiral supermultiplet. The spin-0 superpartners of the SM fermions are appended with an *s*; for example, the electron superpartner is named selectron. Additionally, the superpartners are denoted with a "tilde"; for example, the left-handed selectron is written $\tilde{e}_L$. Note that the subscript on the selectron refers to the handedness of the SM partner of the selectron since the selectrons are spin-0. The left-handed and right-handed components of the fermions are separate two-component Weyl fermions with different gauge transformations. Therefore, they must have separate complex scalar partners. The gauge interactions of the scalar partners are the same as the corresponding SM fermion.

The Higgs boson must reside in a chiral multiplet since it is spin-0. Actually, it turns out that we need two multiplets. This can be simply understood from the fact that the superpotential is holomorphic, so the Higgs multiplet giving mass to the up-type quarks could not give mass to the down-type quarks. So to have masses for all the quarks, we need to have at least two Higgs fields, one with $Y=\frac{1}{2}$ and the other with $Y=-\frac{1}{2}$. Another reason for needing two multiplets is that, if we just had one, then the electroweak gauge symmetry would suffer a triangle gauge anamoly. This lists all the chiral supermultiplets needed for MSSM and are summarized in Table 3.1.

The vector bosons of the SM reside in gauge supermultiplets. Their fermionic superpartners are refered to as gauginos. The color interaction of QCD is mediated via gluons, and their superpartners are called gluinos. As before, a tilde is used to denote the superpartner. Table 3.2 summarizes the gauge supermultiplets of MSSM.

In Tables 3.1, 3.2, the dotted and undotted indices, α, $\dot{\alpha}$, indicate two-component Weyl spinor fields. In the subsequent analysis, we will recast all the spin $\frac{1}{2}$ fields as four-component Dirac spinor fields, which will be represented using the same symbols, but without the dotted and undotted "α"s. For example, $\nu_{L\alpha}$ is the Weyl representation of the left-handed neutrino field, while $\nu_L = \begin{pmatrix} \nu_{L\alpha} \\ \bar{\nu}_L^{\dot{\alpha}} \end{pmatrix}$ is the Dirac field.

Based on $SU(3)\times SU(2)\times U(1)$ symmetry and supersymmetry and R-parity, we could construct the renormalizable Lagrangian of the MSSM as:

$$I = I_{YM} + I_{\text{Kin}} + I_{\text{Matter}} + I_{\text{Breaking}}, \tag{3.4}$$

where

$$I_{\text{Kin}} = \int dx^4 d\theta^2 d\bar{\theta}^2 \left[\overline{Q} e^{2V_Q} Q + U^c e^{-2V_U} \overline{U^c} + D^c e^{-2V_D} \overline{D^c} + \overline{L} e^{2V_L} L + E^c e^{-2V_E} \overline{E^c} \right.$$
$$\left. + \overline{H_B} e^{2V_{H_B}} H_B + H_T^c e^{-2V_{H_T}} \overline{H_T^c}\right] \tag{3.5}$$

Table 3.1 Chiral supermultiplets of the MSSM

Name		Spin 0	Spin $\frac{1}{2}$	$SU(3)_C, SU(2)_L, U(1)_Y$
squarks	Q	$(\tilde{u}_L\ \tilde{d}_L)$	$(u_{L\alpha}\ d_{L\alpha})$	$(3,2,\frac{1}{6})$
and quarks	U^c	$\tilde{u}_R^*$	$\bar{u}_R^{\dot{\alpha}}$	$(\bar{3},1,-\frac{2}{3})$
(×3 families)	D^c	$\tilde{d}_R^*$	$\bar{d}_R^{\dot{\alpha}}$	$(\bar{3},1,\frac{1}{3})$
sleptons,	L	$(\tilde{\nu}_L\ \tilde{e}_L)$	$(\nu_{L\alpha}\ e_{L\alpha})$	$(1,2,-\frac{1}{2})$
leptons (×3 families)	E^c	$\tilde{e}_R^*$	$\bar{e}_R^{\dot{\alpha}}$	$(1,1,1)$
Higgs,	H_T	$(h_T^+\ h_T^0)$	$(\tilde{h}_{T\alpha}^+\ \tilde{h}_{T\alpha}^0)$	$(1,2,\frac{1}{2})$
higgsinos	H_B	$(h_B^0\ h_B^-)$	$(\tilde{h}_{B\alpha}^0\ \tilde{h}_{B\alpha}^-)$	$(1,2,-\frac{1}{2})$

Table 3.2 Gauge supermultiplets in the MSSM

Names	Spin $\frac{1}{2}$	Spin 1	$SU(3)_C, SU(2)_L, U(1)_Y$
Gluino, gluon	$\tilde{g}_\alpha$	g_μ	$(8,1,0)$
Winos, W bosons	$\tilde{W}^\pm_\alpha\ \tilde{W}^0_\alpha$	$W^\pm_\mu\ W^0_\mu$	(1, 3, 0)
Bino, B boson	$\tilde{B}_\alpha$	B_μ	$(1,1,0)$

$$I_{\text{Matter}} = \int dx^4 d\theta^2 \Big[-mH_{Ta}H_B^a + g_E^{ff'} E_f^c \epsilon_{ab} L_{f'}^a H_B^b + g_D^{ff'} D_f^c \epsilon_{ab} Q_{f'}^a H_B^b + g_U^{ff'} U_f^c H_{Ta}^c Q_{f'}^a + H.C. \Big] \tag{3.6}$$

All the above equations are expressed in superfields. $Q, U^c, D^c, L, E^c, H_T, H_B$ are all chiral superfields and $\bar{Q}, \bar{U}^c, \bar{D}^c, \bar{L}, \bar{E}^c, \bar{H}_T, \bar{H}_B$ are anti-chiral superfields. Chiral superfields and anti-chiral superfields have component structures as follows:

$$E = e^{i\theta\sigma^\mu\bar{\theta}\partial_\mu}[\tilde{e}(x) + \sqrt{2}\theta^\alpha e_\alpha(x) + \theta^2 F_E(x)] \tag{3.7}$$

$$E^c = e^{i\theta\sigma^\mu\bar{\theta}\partial_\mu}[\tilde{e}^c(x) + \sqrt{2}\theta^\alpha e^c_\alpha(x) + \theta^2 F_{E^c}(x)] \tag{3.8}$$

$$\bar{E} = e^{-i\theta\sigma^\mu\bar{\theta}\partial_\mu}[\tilde{e}^*(x) + \sqrt{2}\bar{\theta}_{\dot{\alpha}}\bar{e}^{\dot{\alpha}}(x) + \theta^2 F^*_E(x)] \tag{3.9}$$

$$\bar{E}^c = e^{-i\theta\sigma^\mu\bar{\theta}\partial_\mu}[\tilde{e^c}^*(x) + \sqrt{2}\bar{\theta}_{\dot{\alpha}}\bar{e^c}^{\dot{\alpha}}(x) + \theta^2 F^*_{E^c}(x)] \tag{3.10}$$

In Eq. (3.5)

$$V_Q = g_3 G^I \cdot \frac{\lambda^I}{2} + g_2 W^A \cdot \frac{\sigma^A}{2} + \frac{1}{6} g_1 Y \tag{3.11}$$

$$V_U = g_3 G^I \cdot \frac{\lambda^I}{2} + \frac{2}{3} g_1 Y \tag{3.12}$$

$$V_D = g_3 G^I \cdot \frac{\lambda^I}{2} - \frac{1}{3} g_1 Y \tag{3.13}$$

$$V_L = g_2 W^A \cdot \frac{\sigma^A}{2} - \frac{1}{2} g_1 Y \tag{3.14}$$

$$V_E = -g_{1Y} \tag{3.15}$$

$$V_{H_d} = g_2 W^A \cdot \frac{\sigma^A}{2} - \frac{1}{2} g_1 Y \tag{3.16}$$

$$V_{H_u} = g_2 W^A \cdot \frac{\sigma^A}{2} - \frac{1}{2} g_1 Y \tag{3.17}$$

where G^I, W^A and Y are vector superfields. In Wess–Zumino gauge, they have the component structure:

$$G^I = -\theta\sigma^\mu\bar{\theta}G^I_\mu + i\theta\theta\bar{\theta}\bar{\lambda}^I_G - i\bar{\theta}\bar{\theta}\theta\lambda^I_G + \frac{1}{2}\theta\theta\bar{\theta}\bar{\theta}D^I_G \tag{3.18}$$

As mentioned earlier, if supersymmetry were an exact symmetry, particles and their SUSY partners would be degenerate in mass and we would have been able to observe the selectrons, photinos and gluinos by now. Since we have not able to do so, we know that SUSY is a broken symmetry. From a theoretical perspective, the symmetry must be spontaneously broken (SB), analogous to the electroweak symmetry in the SM. Many models of spontaneous symmetry breaking have indeed been proposed. These always involve extending the MSSM to include new particles and interactions at very high mass scales, and there is no consensus on exactly how this should be done. However, from a practical point of view, it is extremely useful to simply parameterize our ignorance of these issues by just introducing extra terms that softly break the supersymmetry explicitly in the effective MSSM Lagrangian. In doing so, one finds the breaking terms:

$$\begin{aligned} l_{\text{Breaking}} = & -\frac{1}{2}(M_3\tilde{g} + M_2\tilde{W}\tilde{W} + M_1\tilde{B}\tilde{B} + c.c.) \\ & - (\tilde{u}_R a_u \tilde{Q} H_u - \tilde{d}_R a_d \tilde{Q} H_d - \tilde{e}_R a_e \tilde{L} H_d + c.c) \\ & - \tilde{Q}^\dagger m^2_Q \tilde{Q} - \tilde{L}^\dagger m^2_L \tilde{L} - \tilde{u}_R m^2_{U^c} \tilde{u}^*_R - \tilde{d}_R m^2_{D^c} \tilde{d}^*_R - \tilde{e}_R m^2_{E^c} \tilde{e}^*_R \\ & - m^2_{H_u} H^*_u H_u - m^2_{H_d} H^*_d H_d - (bH_u H_d + c.c.). \end{aligned} \tag{3.19}$$

References

S. Coleman, and J. Mandula, Phys. Rev. **159** (1967) 1251

E. Gildener, Phys. Rev. D **14**, 1667 (1976)

R. Haag, J. Lopuszańki, and M. Sohnius, Nucl. Phys. B **88**, 257 (1975)

G. Jungman, M. Kamionkowski, and K. Griest, Phys. Rep. **267**, 195 (1996)

S.P. Martin, A supersymmetry Primer (1997).

L. Susskind, Phys. Rev. D **20**, 2619 (1979)

G. t Hooft, in Recent developments in gauge theories, Proceedings of the NATO Advanced Summer Institute, Cargese 1979 (Plenum, 1980)

S. Weinberg, Phys. Rev. D **13**, 974 (1976)

The ALEPH, DELPHI, L3, OPAL, SLD Collaborations, the LEP Electroweak Working Group, the SLD Electroweak and Heavy Flavour Groups, Phys. Rep. **427**, 257 (2006), hep-ex/0509008

Chapter 4
The Muonium–Antimuonium Oscillation in the Extended Minimal Supersymmetric Standard Model

4.1 The Extended Minimal Supersymmetric Standard Model

In order to implement the see–saw mechanism (Minkowski 1977; Yanagida 1979; Gell-Mann et al. 1979; Glashow 1980; Mohapatra and Senjanovic 1980) for neutrino masses, we consider an extension of the MSSM, where one adds three additional gauge singlet chiral superfields N_i^c (i $= e, \mu, \tau$ denotes the generation), whose θ-component is a right-handed neutrino field,

$$N_i^c = \tilde{\nu}_{iR}^*(y) + \sqrt{2}\theta^\alpha \nu_R(y)_\alpha + \theta^\alpha\theta_\alpha F_{N_i^c}(y), \tag{4.1}$$

where

$$y^\mu = x^\mu + i\theta^\alpha \sigma^\mu_{\alpha\dot{\alpha}}\bar{\theta}^{\dot{\alpha}}. \tag{4.2}$$

These $SU(3) \times SU(2)_L \times U(1)$ singlet superfields are coupled to other MSSM superfields via the superpotential. We employ the most general R-parity conserving renormalizable superpotential so that the superpotential is

$$W = -\mu\epsilon_{ab}H_B^a H_T^b + \lambda_i\epsilon_{ab}E_i^c L_i^a H_B^b + \lambda_i'\epsilon_{ab}H_T^a L_i^b N_i^c + \frac{1}{2}M_R^i N_i^c N_i^c, \tag{4.3}$$

while the relevant soft supersymmetry breaking terms are

$$\begin{aligned}\mathcal{L}_{\text{soft}}^{\text{EMSSM}} &= -(m_{\tilde{L}}^i)^2\left(\tilde{\nu}_{iL}^*\tilde{\nu}_{iL} + \tilde{\ell}_{iL}^*\tilde{\ell}_{iL}\right) - (m_{\tilde{R}}^i)^2\tilde{\ell}_{iR}^*\tilde{\ell}_{iR} - (m_{\tilde{N}}^i)^2\tilde{\nu}_{iR}^*\tilde{\nu}_{iR} \\ &\quad - \left(\lambda_i' A_i h_T^0\tilde{\nu}_{iL}\tilde{\nu}_{iR}^* + M_R^i B_i\tilde{\nu}_{iR}\tilde{\nu}_{iR} + \lambda_i C_i h_B^0\tilde{\ell}_{iL}\tilde{\ell}_{iR}^* + H.C.\right).\end{aligned} \tag{4.4}$$

The interaction terms that contribute to the muonium–antimuonium oscillation and the electron and muon anomalous magnetic moments can be extracted from the Lagrangian of this extended Minimal Supersymmetric Standard Model (EMSSM) as

$$\mathcal{L}_{\text{int}}^W = -\frac{g_2}{\sqrt{2}}\left(W^{-\mu}\bar{\ell}_{iL}\gamma_\mu\nu_{iL} + W^{+\mu}\bar{\nu}_{iL}\gamma_\mu\ell_{iL}\right), \tag{4.5}$$

B. Liu, *Muonium–Antimuonium Oscillations in an Extended Minimal Supersymmetric Standard Model*, Springer Theses,
DOI: 10.1007/978-1-4419-8330-5_4,

$$\mathcal{L}_{\text{int}}^{\tilde{W}^-} = -ig_2\Big(\bar{\ell}_{iL}\,\tilde{W}^-\,\tilde{\nu}_{iL} - \tilde{\nu}_{iL}^*\,\overline{\tilde{W}^-}\,\ell_{iL}\Big), \tag{4.6}$$

$$\mathcal{L}_{\text{int}}^{\tilde{W}^0} = \frac{g_2 i}{\sqrt{2}}\Big(\bar{\ell}_{iL}\,\tilde{W}^0\,\tilde{\ell}_{iL} - \tilde{\ell}_{iL}^*\,\overline{\tilde{W}^0}\,\ell_{iL}\Big), \tag{4.7}$$

$$\mathcal{L}_{\text{int}}^{\tilde{B}} = \frac{g_1 i}{\sqrt{2}}\Big(\bar{\ell}_{iL}\,\tilde{B}\,\tilde{\ell}_{iL} - \tilde{\ell}_{iL}^*\,\overline{\tilde{B}}\,\ell_{iL}\Big) + \sqrt{2}g_1 i\Big(\bar{\ell}_{iR}\,\tilde{B}\,\tilde{\ell}_{iR} - \tilde{\ell}_{iR}^*\,\overline{\tilde{B}}\,\ell_{iR}\Big), \tag{4.8}$$

$$\mathcal{L}_{\text{int}}^{\tilde{h}_{\bar{B}}^-} = \frac{m_i}{V_B}\Big(\bar{\ell}_{iR}\,\tilde{h}_{\bar{B}}^-\,\tilde{\nu}_{iL} + \overline{\tilde{h}_{\bar{B}}^-}\,\ell_{iR}\,\tilde{\nu}_{iL}^*\Big) + \frac{m_D^i}{V_T}\Big(\bar{\ell}_{iL}\,\tilde{h}_{\bar{B}}^-\,\tilde{\nu}_{iR} + \overline{\tilde{h}_{\bar{B}}^-}\,\ell_{iL}\,\tilde{\nu}_{iR}^*\Big), \tag{4.9}$$

$$\mathcal{L}_{\text{int}}^{\tilde{h}_B^0} = -\frac{m_i}{V_B}\Big(\bar{\ell}_{iL}\,\tilde{h}_B^0\,\tilde{\ell}_{iR} + \tilde{\ell}_{iR}^*\,\overline{\tilde{h}_B^0}\,\ell_{iL}\Big) - \frac{m_i}{V_B}\Big(\bar{\ell}_{iR}\,\tilde{h}_B^0\,\tilde{\ell}_{iL} + \tilde{\ell}_{iL}^*\,\overline{\tilde{h}_B^0}\,\ell_{iR}\Big). \tag{4.10}$$

In the above equations, all the spin $\frac{1}{2}$ fields are Dirac spinor fields. In particular, note that the field $\tilde{h}_{\bar{B}}^-$ has the Weyl field decomposition

$$\tilde{h}_{\bar{B}}^- = \begin{pmatrix} \tilde{h}_{\bar{B}\alpha}^- \\ \bar{\tilde{h}}_T^{+\dot{\alpha}} \end{pmatrix}. \tag{4.11}$$

The parameters V_B and V_T are the vacuum expectation values of the two Higgs fields: $\langle h_{\bar{B}}^0\rangle = V_B$ and $\langle h_T^0\rangle = V_T$. These VEVs are related to the known mass of the W boson and the electroweak gauge couplings as

$$V_B^2 + V_T^2 = V^2 = \frac{2M_W^2}{g_2^2} \approx (174\,\text{GeV})^2, \tag{4.12}$$

while the ratio of the VEVs is traditionally written as

$$\tan\beta \equiv \frac{V_T}{V_B}. \tag{4.13}$$

In the above, $m_D^i = \lambda_i' V_T$ are the Dirac mass parameters of neutrinos and m_i are the lepton masses. Since the masses of electron and muon are small, the terms which have couplings proportional to m_i/V_B in interactions (4.9) and (4.10) are severely suppressed and will be ignored in the subsequent analysis.

4.2 Neutrino and Sneutrino Mass Eigenstates in the EMSSM

The neutrino mass term can be extracted from the superpotential terms $\lambda_i'\epsilon_{ab}H_T^aL_i^bN_i^c$ and $\frac{1}{2}M_R^iN_i^cN_i^c$ in Eq. 4.3 as

$$\mathcal{L}_{\text{mass}}^{\nu_i} = -\frac{1}{2}\begin{pmatrix}\overline{(\nu_{iL})^c} & \overline{\nu_{iR}}\end{pmatrix}\begin{pmatrix} 0 & m_D^i \\ m_D^i & M_R^i \end{pmatrix}\begin{pmatrix} \nu_{iL} \\ (\nu_{iR})^c \end{pmatrix} + H.C. \tag{4.14}$$

Note that the upper left element in the neutrino mass matrix is zero. This element involves only left-handed neutrinos and in our EMSSM its generation requires a nonrenormalizable superpotential term. Consequently, we ignore this term. For every generation, the two mass eigenvalues, $m_a^{\nu_i}$, are obtained from the diagonalization of the 2×2 matrix

$$M^{\nu_i} = \begin{pmatrix} 0 & m_D^i \\ m_D^i & M_R^i \end{pmatrix}. \tag{4.15}$$

Since M^{ν_i} is symmetric, it can be diagonalized by a single unitary 2×2 matrix, V^i, as

$$M_{\text{diag}}^{\tilde{\nu}_i} = V^{iT} M^{\tilde{\nu}_i} V^i. \tag{4.16}$$

This diagonalization is implemented via the basis change as following

$$\begin{pmatrix} \nu_{iL} \\ (\nu_{iR})^c \end{pmatrix} = V^i \begin{pmatrix} \nu_{i1} \\ (\nu_{i2})^c \end{pmatrix}, \quad \begin{pmatrix} (\nu_{iL})^c \\ \nu_{iR} \end{pmatrix} = V^{i*} \begin{pmatrix} (\nu_{i1})^c \\ \nu_{i2} \end{pmatrix}. \tag{4.17}$$

The neutrino mass term then takes the form

$$\mathcal{L}_{\text{mass}}^{\nu_i} = -\frac{1}{2}\sum_{a=1}^{2} m_A^{\nu_i}\left[\nu_{ia}^T C \nu_{ia} + \overline{\nu_{ia}} C \overline{\nu_{ia}^T}\right] = -\sum_{a=1}^{2} m_a^{\nu_i} \overline{\nu_{ia}} \nu_{ia}, \tag{4.18}$$

where $m_A^{\nu_i}$ are the Majorana neutrino masses.

Since a nonzero Majorana mass parameter M_R^i does not require $SU(2)_L \times U(1)$ symmetry breaking, it is naturally much bigger than the Dirac mass parameter m_D^i whose nontrivial value does require $SU(2)_L \times U(1)$ symmetry breaking. So doing, one finds on diagonalization of the 2×2 neutrino mass matrix that the two eigenvalues are crudely given by

$$m_1^{\nu_i} \sim \frac{(m_D^i)^2}{M_R^i} \ll m_D^i, \quad m_2^{\nu_i} \sim M_R^i. \tag{4.19}$$

This constitutes the so called see–saw mechanism (Minkowski 1977; Yanagida 1979; Gell-Mann et al. 1979; Glashow 1980; Mohapatra and Senjanovic 1980) and provides a natural explanation of the smallness of the three light neutrino masses. Moreover, the elements of the mixing matrix are characterized by an m_D^i/M_R^i dependence. We expand V^i in power series of the matrix parameter $\xi_i = \frac{m_D^i}{M_R^i}$, with the constraint $\xi_i \ll 1$. The form of V^i to first order in ξ_i can be estimated to be

$$V^i = \begin{pmatrix} 1 & \xi_i \\ -\xi_i & 1 \end{pmatrix}. \tag{4.20}$$

The sneutrino masses are obtained by diagonalizing a 4×4 squared mass matrix. Here, it is convenient to define $\tilde{\nu}_{iL} = \frac{1}{\sqrt{2}}(\tilde{\nu}_{iL1} + i\tilde{\nu}_{iL2})$ and $\tilde{\nu}_{iR} = \frac{1}{\sqrt{2}}(\tilde{\nu}_{iR1} +$

$i\tilde{\nu}_{iR2})$. Then, the sneutrino-squared mass matrix separates into CP-even and CP-odd blocks (Yuval and Howard 1997),

$$\begin{aligned}\mathcal{L}^{\tilde{\nu}_i}_{\text{mass}} &= \frac{1}{2}\begin{pmatrix}\phi^i_1 & \phi^i_2\end{pmatrix}\mathcal{M}^2_{\tilde{\nu}_i}\begin{pmatrix}\phi^i_1 \\ \phi^i_2\end{pmatrix} \\ &= \frac{1}{2}\begin{pmatrix}\phi^i_1 & \phi^i_2\end{pmatrix}\begin{pmatrix}\mathcal{M}^2_{\tilde{\nu}_i+} & 0 \\ 0 & \mathcal{M}^2_{\tilde{\nu}_i-}\end{pmatrix}\begin{pmatrix}\phi^i_1 \\ \phi^i_2\end{pmatrix},\end{aligned} \tag{4.21}$$

where $\phi^i_a \equiv (\tilde{\nu}_{iLa} \quad \tilde{\nu}_{iRa})$ and $\mathcal{M}^2_{\tilde{\nu}_i\pm}$ consist of the following 2×2 blocks:

$$\mathcal{M}^2_{\tilde{\nu}_i\pm} = \begin{pmatrix}(m^i_{\tilde{L}})^2 + \frac{1}{2}m^2_Z\cos 2\beta + (m^i_D)^2 & m^i_D(A_i - \mu\cot\beta \pm M^i_R) \\ m^i_D(A_i - \mu\cot\beta \pm M^i_R) & (M^i_R)^2 + (m^i_D)^2 + (m^i_{\tilde{N}})^2 \pm 2B_iM^i_R\end{pmatrix}, \tag{4.22}$$

with A_i and B_i are SUSY breaking parameters (cf. Eq. 4.4). Since the sneutrino mass matrix $\mathcal{M}^2_{\tilde{\nu}_i}$ is real and symmetric, it can be diagonalized by a real orthogonal 4×4 matrix, U^i, as

$$\mathcal{M}^2_{\tilde{\nu}_i\text{diag}} = U^{iT}\mathcal{M}^2_{\tilde{\nu}_i}U^i, \tag{4.23}$$

where U^i is in a form as

$$U^i = \begin{pmatrix}U^i_+ & 0 \\ 0 & U^i_-\end{pmatrix}. \tag{4.24}$$

This diagonalization is implemented via the basis change on ϕ^i_1 and ϕ^i_2

$$\begin{pmatrix}\phi^i_1 \\ \phi^i_2\end{pmatrix} = \begin{pmatrix}\tilde{\nu}_{iL1} \\ \tilde{\nu}_{iR1} \\ \tilde{\nu}_{iL2} \\ \tilde{\nu}_{iR2}\end{pmatrix} = U^i\begin{pmatrix}\tilde{\nu}_{i1} \\ \tilde{\nu}_{i2} \\ \tilde{\nu}_{i3} \\ \tilde{\nu}_{i4}\end{pmatrix}, \tag{4.25}$$

where $\tilde{\nu}_{ia}$ are all real. Then the sneutrino mass term takes the form

$$\mathcal{L}^{\tilde{\nu}_i}_{\text{mass}} = -\frac{1}{2}\sum_{a=1}^{4} m^{\tilde{\nu}_i}_a\tilde{\nu}_{ia}\tilde{\nu}_{ia}, \tag{4.26}$$

where $m^{\tilde{\nu}_i}_a$ are the sneutrino mass eigenvalues.

In the following derivation we assume that M^i_R is the largest mass parameter. Then, to the first order in $1/M^i_R$, the two light mass eigenvalues are roughly

$$\begin{aligned}m^2_{\tilde{\nu}_{i1}} &\approx (m^i_{\tilde{L}})^2 + \frac{1}{2}m^2_Z\cos 2\beta - \frac{2(m^i_D)^2(A_i - \mu\cot\beta - B_i)}{M^i_R}, \\ m^2_{\tilde{\nu}_{i3}} &\approx (m^i_{\tilde{L}})^2 + \frac{1}{2}m^2_Z\cos 2\beta + \frac{2(m^i_D)^2(A_i - \mu\cot\beta - B_i)}{M^i_R},\end{aligned} \tag{4.27}$$

while the two heavy mass eigenvalues are

$$\begin{aligned} m^2_{\tilde{\nu}_{i2}} &\approx (M^i_R)^2 + 2B_i M^i_R, \\ m^2_{\tilde{\nu}_{i4}} &\approx (M^i_R)^2 - 2B_i M^i_R. \end{aligned} \tag{4.28}$$

To avoid excessive complication in our calculations, we expand U^i in powers of the matrix parameter $\xi_i = \frac{m^i_D}{M^i_R}$. The form of U to first order of ξ_i is

$$U^i = \begin{pmatrix} U^i_+ & 0 \\ 0 & U^i_- \end{pmatrix} = \begin{pmatrix} \begin{pmatrix} 1 & \xi_i \\ -\xi_i & 1 \end{pmatrix} & 0 \\ 0 & \begin{pmatrix} 1 & -\xi_i \\ \xi_i & 1 \end{pmatrix} \end{pmatrix}. \tag{4.29}$$

For every generation, the slepton mass term is given by

$$\mathcal{L}^{\tilde{\ell}_i}_{\text{mass}} = \begin{pmatrix} \tilde{\ell}^*_{iL} & \tilde{\ell}^*_{iR} \end{pmatrix} \begin{pmatrix} (m^{LL}_{\tilde{\ell}_i})^2 & (m^{LR}_{\tilde{\ell}_i})^2 \\ (m^{LR}_{\tilde{\ell}_i})^2 & (m^{RR}_{\tilde{\ell}_i})^2 \end{pmatrix} \begin{pmatrix} \tilde{\ell}_{iL} \\ \tilde{\ell}_{iR} \end{pmatrix}, \tag{4.30}$$

where

$$(m^{LL}_{\tilde{\ell}_i})^2 = (m^i_{\tilde{L}})^2 + m^2_Z \cos 2\beta \left(\sin^2\theta_W - \frac{1}{2} \right), \tag{4.31}$$

$$(m^{LR}_{\tilde{\ell}_i})^2 = \lambda_i \mu V_T + \lambda_i C_i V_B, \tag{4.32}$$

$$(m^{RR}_{\tilde{\ell}_i})^2 = (m^i_{\tilde{R}})^2 - m^2_Z \cos 2\beta \sin^2\theta_W. \tag{4.33}$$

Since $\lambda_i V_B = m_i$, the off diagonal matrix element, $(m^{LR}_{\tilde{\ell}_i})^2$, can be written as

$$(m^{LR}_{\tilde{\ell}_i})^2 = m_i(\mu \tan\beta + C_i). \tag{4.34}$$

Because the masses of electron and muon are very small compared with the sparticle mass scale, we ignore these off diagonal terms and consider $\tilde{\ell}_{iL}$ and $\tilde{\ell}_{iR}$ as mass eigenstates.

Inserting the transformation (4.17) and (4.25) in the interaction terms (4.5–4.9) yields the explicit interactions in their mass basis:

$$\mathcal{L}^W_{\text{int}} = -\frac{g_2}{\sqrt{2}} \sum_{a=1}^{2} \left(W^{-\mu} \bar{\ell}_{iL} \gamma_\mu V^i_{1a} \nu_{ia} + W^{+\mu} \bar{\nu}_{ia} V^{i*}_{1a} \gamma_\mu \ell_{iL} \right), \tag{4.35}$$

$$\mathcal{L}^{\tilde{W}^-}_{\text{int}} = -\frac{ig_2}{\sqrt{2}} \sum_{a=1}^{2} \bar{\ell}_{iL}\, \tilde{W}^-\, U^i_{1a}\, \tilde{\nu}_{ia} + \frac{g_2}{\sqrt{2}} \sum_{a=3}^{4} \bar{\ell}_{iL}\, \tilde{W}^-\, U^i_{3a}\, \tilde{\nu}_{ia} + H.C., \tag{4.36}$$

$$\mathcal{L}^{\tilde{W}^0}_{\text{int}} = \frac{g_2 i}{\sqrt{2}} \left(\bar{\ell}_{iL} \tilde{W}^0 \tilde{\ell}_{iL} - \tilde{\ell}^*_{iL}\, \overline{\tilde{W}^0}\, \ell_{iL} \right), \tag{4.37}$$

$$\mathcal{L}_{\text{int}}^{\tilde{B}} = \frac{g_1 i}{\sqrt{2}}\left(\bar{\ell}_{iL}\,\tilde{B}\,\tilde{\ell}_{iL} - \tilde{\ell}_{iL}^{*}\,\bar{\tilde{B}}\,\ell_{iL}\right) + \sqrt{2}g_1 i\left(\bar{\ell}_{iR}\,\tilde{B}\,\tilde{\ell}_{iR} - \tilde{\ell}_{iR}^{*}\,\bar{\tilde{B}}\,\ell_{iR}\right), \tag{4.38}$$

$$\mathcal{L}_{\text{int}}^{\tilde{h}_B^-} = \frac{m_D^i}{\sqrt{2}V_T}\sum_{a=1}^{2}\bar{\ell}_{iL}\,\tilde{h}_B^-\,U_{2a}^i\tilde{\nu}_{ia} + \frac{i m_D^i}{\sqrt{2}V_T}\sum_{a=3}^{4}\bar{\ell}_{iL}\,\tilde{h}_B^-\,U_{4a}^i\tilde{\nu}_{ia} + H.C. \tag{4.39}$$

4.3 The Muonium–Antimuonium Oscillation in the EMSSM

The lowest order Feynman diagrams accounting for muonium and antimuonium mixing are displayed in Fig. 4.1. Graphs (a) and (b) are the non-SUSY contributions, which are mediated by Majorana neutrinos and W boson. The other graphs all involve SUSY partners. Graphs (c) and (d) are mediated by sneutrinos and wino, while graph (e) and (f) are mediated by sneutrinos and higgsino.

The T-matrix elements of graphs (a) and (b) are

$$\begin{aligned} T_a = T_b = &\frac{g_2^4}{512\pi^2 M_W^2}[\bar{\mu}(3)\gamma^\mu(1-\gamma_5)e(2)][\bar{\mu}(4)\gamma_\mu(1-\gamma_5)e(1)] \\ &\times\sum_{a=1}^{2}\sum_{b=1}^{2}(V_{1a}^{\mu})^2(V_{1b}^{e*})^2K(x_{\nu_{\mu a}}, x_{\nu_{eb}}), \end{aligned} \tag{4.40}$$

where $\bar{\mu}(3) = \bar{\mu}(p_3, s_3), \bar{\mu}(4) = \bar{\mu}(p_4, s_4), e(1) = e(p_1, s_1)$ and $e(2) = e(p_2, s_2)$ are the spinors of the muons and electrons and $x_{\nu_{ia}} = \frac{m_{\nu_{ia}}^2}{M_W^2}$, $a = 1, 2$. The function $K(x_{\nu_{\mu a}}, x_{\nu_{eb}})$ takes the form

$$K(x_A, x_B) = \sqrt{x_A x_B}\frac{L(x_A, x_B) - L(x_B, x_A)}{x_A - x_B}, \tag{4.41}$$

with

$$L(x_A, x_B) = \frac{4 - x_A x_B}{(x_A - 1)} + \frac{x_A(2x_B - x_A x_B - 4)}{(x_A - 1)^2}\ln x_A. \tag{4.42}$$

The T-matrix elements of graphs (c) and (d) are

$$\begin{aligned} T_c = T_d = &-\frac{g_2^4}{1024\pi^2 M_{\tilde{W}^-}^2}[\bar{\mu}(3)\gamma^\mu(1-\gamma_5)e(2)][\bar{\mu}(4)\gamma_\mu(1-\gamma_5)e(1)] \\ &\times\Bigg(\sum_{a=1}^{2}\sum_{b=1}^{2}(U_{1a}^{\mu})^2(U_{1b}^{e})^2 I(y_{\tilde{\nu}_{\mu a}}, y_{\tilde{\nu}_{eb}}) - \sum_{a=1}^{2}\sum_{b=3}^{4}(U_{1a}^{\mu})^2(U_{3b}^{e})^2 I(y_{\tilde{\nu}_{\mu a}}, y_{\tilde{\nu}_{eb}}) \\ &-\sum_{a=3}^{4}\sum_{b=1}^{2}(U_{3a}^{\mu})^2(U_{1b}^{e})^2 I(y_{\tilde{\nu}_{\mu a}}, y_{\tilde{\nu}_{eb}}) + \sum_{a=3}^{4}\sum_{b=3}^{4}(U_{3a}^{\mu})^2(U_{3b}^{e})^2 I(y_{\tilde{\nu}_{\mu a}}, y_{\tilde{\nu}_{eb}})\Bigg), \end{aligned} \tag{4.43}$$

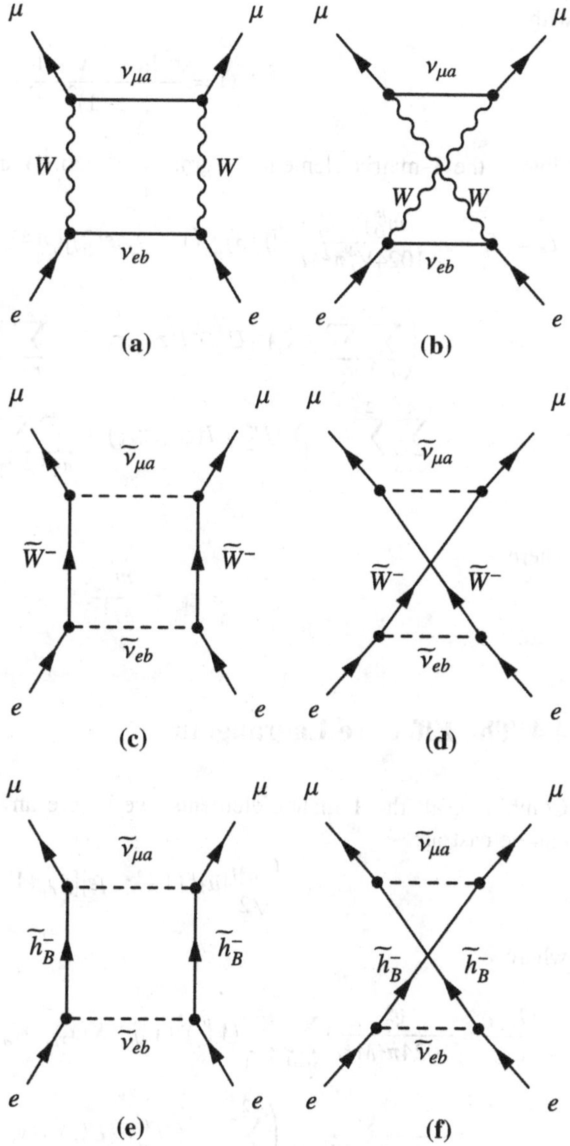

Fig. 4.1 Feynman graphs contributing to the muonium–antimuonium mixing

where

$$y_{\tilde{\nu}_{ia}} = \frac{m^2_{\tilde{\nu}_{ia}}}{M^2_{\tilde{W}^-}}, \tag{4.44}$$

$$I(x_1, x_2) = \frac{J(x_1) - J(x_2)}{x_1 - x_2}, \tag{4.45}$$

with

$$J(x) = \frac{x^2 \ln x - x + 1}{(x-1)^2}. \tag{4.46}$$

Finally, the T-matrix elements of graphs (e) and (f) are

$$\begin{aligned} T_e = T_f = &-\frac{(m_D^{\mu})^2 (m_D^{e})^2}{1024 V_T^4 \pi^2 M_{\tilde{h}_B^-}^2} [\bar{\mu}(3)\gamma^{\mu}(1-\gamma_5)e(2)][\bar{\mu}(4)\gamma_{\mu}(1-\gamma_5)e(1)] \\ &\times \left(\sum_{a=1}^{2}\sum_{b=1}^{2} (U_{2a}^{\mu})^2 (U_{2b}^{e})^2 I(z_{\tilde{\nu}_{\mu a}}, z_{\tilde{\nu}_{eb}}) - \sum_{a=1}^{2}\sum_{b=3}^{4} (U_{2a}^{\mu})^2 (U_{4b}^{e})^2 I(z_{\tilde{\nu}_{\mu a}}, z_{\tilde{\nu}_{eb}}) \right. \\ &\left. - \sum_{a=3}^{4}\sum_{b=1}^{2} (U_{4a}^{\mu})^2 (U_{2b}^{e})^2 I(z_{\tilde{\nu}_{\mu a}}, z_{\tilde{\nu}_{eb}}) + \sum_{a=3}^{4}\sum_{b=3}^{4} (U_{4a}^{\mu})^2 (U_{4b}^{e})^2 I(z_{\tilde{\nu}_{\mu a}}, z_{\tilde{\nu}_{eb}}) \right), \end{aligned} \tag{4.47}$$

where

$$z_{\tilde{\nu}_{ia}} = \frac{m_{\tilde{\nu}_{ia}}^2}{M_{\tilde{h}_B^-}^2}. \tag{4.48}$$

4.4 The Effective Lagrangian

Combining all the T-matrix elements, we secure an effective Lagrangian which can be cast as:

$$\mathcal{L}_{\text{eff}} = \frac{G_{\bar{M}M}}{\sqrt{2}} [\bar{\mu}\gamma^{\mu}(1-\gamma_5)e][\bar{\mu}\gamma_{\mu}(1-\gamma_5)e], \tag{4.49}$$

where

$$\begin{aligned} \frac{G_{\bar{M}M}}{\sqrt{2}} = &\frac{g_2^4}{1024\pi^2 M_W^2} \cdot \sum_{a=1}^{2}\sum_{b=1}^{2} (V_{1a}^{\mu})^2 (V_{1b}^{e*})^2 K(x_{\nu_{\mu a}}, x_{\nu_{eb}}) \\ &- \frac{g_2^4}{2048\pi^2 M_{\tilde{W}^-}^2} \cdot \left(\sum_{a=1}^{2}\sum_{b=1}^{2} (U_{1a}^{\mu})^2 (U_{1b}^{e})^2 I(y_{\tilde{\nu}_{\mu a}}, y_{\tilde{\nu}_{eb}}) \right. \\ &- \sum_{a=1}^{2}\sum_{b=3}^{4} (U_{1a}^{\mu})^2 (U_{3b}^{e})^2 I(y_{\tilde{\nu}_{\mu a}}, y_{\tilde{\nu}_{eb}}) - \sum_{a=3}^{4}\sum_{b=1}^{2} (U_{3a}^{\mu})^2 (U_{1b}^{e})^2 I(y_{\tilde{\nu}_{\mu a}}, y_{\tilde{\nu}_{eb}}). \\ &\left. + \sum_{a=3}^{4}\sum_{b=3}^{4} (U_{3a}^{\mu})^2 (U_{3b}^{e})^2 I(y_{\tilde{\nu}_{\mu a}}, y_{\tilde{\nu}_{eb}}) \right) - \frac{(m_D^{\mu})^2 (m_D^{e})^2}{2048 V_T^4 \pi^2 M_{\tilde{h}_B^-}^2} \end{aligned}$$

$$\cdot\left(\sum_{a=1}^{2}\sum_{b=1}^{2}(U^{\mu}_{2a})^2(U^{e}_{2b})^2 I(z_{\tilde{\nu}_{\mu a}}, z_{\tilde{\nu}_{eb}}) - \sum_{a=1}^{2}\sum_{b=3}^{4}(U^{\mu}_{2a})^2(U^{e}_{4b})^2 I(z_{\tilde{\nu}_{\mu a}}, z_{\tilde{\nu}_{eb}})\right.$$
$$\left. - \sum_{a=3}^{4}\sum_{b=1}^{2}(U^{\mu}_{4a})^2(U^{e}_{2b})^2 I(z_{\tilde{\nu}_{\mu a}}, z_{\tilde{\nu}_{eb}}) + \sum_{a=3}^{4}\sum_{b=3}^{4}(U^{\mu}_{4a})^2(U^{e}_{4b})^2 I(z_{\tilde{\nu}_{\mu a}}, z_{\tilde{\nu}_{eb}})\right). \tag{4.50}$$

4.5 Estimate of the Effective Coupling Constant

The present experimental limit (Willmann 1999) on the non-observation of muonium–antimuonium oscillation translates into the bound

$$|\mathrm{Re}G_{\bar{M}M}| \leq 3.0 \times 10^{-3} G_F, \tag{4.51}$$

where $G_F \simeq 1.16 \times 10^{-5}\mathrm{GeV}^{-2}$ is the Fermi scale. This limit can then be used to construct some constraints on the parameters of this model.

For simplicity, we set the neutrino Dirac masses m^e_D, m^{μ}_D and the right-handed neutrino masses M^e_R, M^{μ}_R to some common mass scales m_D and M_R respectively. The light neutrino mass scale m_ν is of order m_D^2/M_R, while the heavy neutrino mass scale is of order M_R.

Using these assumptions and taking into account the mixing matrices approximations Eq. 4.20 and 4.29, we can simplify the effective coupling constant (4.50) to a more manageable approximated form. The contribution from graphs (a) and (b) in $G_{\bar{M}M}$ is $\frac{g_2^4}{1024\pi^2 M_W^2}\cdot\sum_{a=1}^{2}\sum_{b=1}^{2}(V^{\mu}_{1a})^2(V^{e*}_{1b})^2\cdot K(x_{\nu_{\mu a}}, x_{\nu_{eb}})$. With the limits of $m_{\nu_\mu 1}, m_{\nu_e 1} \sim \mathcal{O}(\frac{m_D^2}{M_R})$ and $m_{\nu_\mu 2}, m_{\nu_e 2} \sim \mathcal{O}(M_R)$, the contribution $(V^{\mu}_{1a})^2(V^{e*}_{1b})^2$ $K(x_{\nu_{\mu a}}, x_{\nu_{eb}})$ can be approximated as

$$\begin{gathered}(V^{\mu}_{11})^2(V^{e*}_{11})^2 K(x_{\nu_{\mu 1}}, x_{\nu_{e1}}) \sim \frac{m_D^4}{M_R^2 M_W^2}\ln\left(\frac{M_R M_W}{m_D^2}\right),\\ (V^{\mu}_{11})^2(V^{e*}_{12})^2 K(x_{\nu_{\mu 1}}, x_{\nu_{e2}}),\quad (V^{\mu}_{12})^2(V^{e*}_{11})^2 K(x_{\nu_{\mu 2}}, x_{\nu_{e2}}) \sim \frac{m_D^8}{M_R^4 M_W^4}\ln\left(\frac{M_R M_W}{m_D^2}\right),\\ (V^{\mu}_{12})^2(V^{e*}_{12})^2 K(x_{\nu_{\mu 2}}, x_{\nu_{e2}}) \sim \frac{m_D^4}{M_R^2 M_W^2}\ln\left(\frac{M_R}{M_W}\right).\end{gathered} \tag{4.52}$$

Taking M_R as the largest mass parameter, the first term and the third term are comparable, while the second one is suppressed by a factor $m_D^4/(M_R^2 M_W^2)$. Therefore, the contribution from graphs (a) and (b) is roughly

$$\frac{g_2^4}{1024\pi^2 M_W^2}\cdot\sum_{a=1}^{2}\sum_{b=1}^{2}(V^{\mu}_{1a})^2(V^{e*}_{1b})^2 K(x_{\nu_{\mu a}}, x_{\nu_{eb}}) \approx \frac{g_2^4 m_D^4}{1024\pi^2 M_R^2 M_W^4}\cdot\ln\frac{M_R}{M_W}. \tag{4.53}$$

The second term in Eq. 4.50 is the contribution of graph (c) and (d), in which the function $I(y_{\tilde{\nu}_{\mu a}}, y_{\tilde{\nu}_{eb}})$ is a decreasing function of $y_{\tilde{\nu}_{\mu a}}$ and $y_{\tilde{\nu}_{eb}}$. It will be small for heavy sneutrinos. To see this, we employ the approximations (4.29)

$$\begin{aligned} &U^{\mu}_{11},\quad U^{e}_{11},\quad U^{\mu}_{33},\quad U^{e}_{33} \sim \mathcal{O}(1),\\ &U^{\mu}_{12},\quad U^{e}_{12},\quad U^{\mu}_{34},\quad U^{e}_{34} \sim \mathcal{O}\left(\frac{m_D}{M_R}\right), \end{aligned} \tag{4.54}$$

so that the terms involving heavy sneutrinos will get an extra suppression from the mixing matrix. Therefore, the contribution of graph (c) and (d) is dominated by the term that only includes the light sneutrinos so that

$$\begin{aligned} &-\frac{g_2^4}{2048\pi^2 M^2_{\tilde{W}^-}}\cdot\left(\sum_{a=1}^{2}\sum_{b=1}^{2}(U^{\mu}_{1a})^2(U^{e}_{1b})^2 I(y_{\tilde{\nu}}, y_{\tilde{\nu}_{eb}}) - \sum_{a=1}^{2}\sum_{b=3}^{4}(U^{\mu}_{1a})^2(U^{e}_{3b})^2 I(y_{\tilde{\nu}}, y_{\tilde{\nu}_{eb}})\right.\\ &\left.-\sum_{a=3}^{4}\sum_{b=1}^{2}(U^{\mu}_{3a})^2(U^{e}_{1b})^2 I(y_{\tilde{\nu}}, y_{\tilde{\nu}_{eb}}) + \sum_{a=3}^{4}\sum_{b=3}^{4}(U^{\mu}_{3a})^2(U^{e}_{3b})^2 I(y_{\tilde{\nu}_{\mu a}}, y_{\tilde{\nu}_{eb}})\right)\\ &\approx -\frac{g_2^4}{2048\pi^2 M^2_{\tilde{W}^-}}\cdot\left(I(y_{\tilde{\nu}_{\mu 1}}, y_{\tilde{\nu}_{e1}}) - I(y_{\tilde{\nu}_{\mu 1}}, y_{\tilde{\nu}_{e3}}) - I(y_{\tilde{\nu}_{\mu 3}}, y_{\tilde{\nu}_{e1}}) + I(y_{\tilde{\nu}_{\mu 3}}, y_{\tilde{\nu}_{e3}})\right). \end{aligned} \tag{4.55}$$

Employing the squared-mass difference between the two light sneutrinos in Eq. 4.27, the above expression can be approximated as

$$\begin{aligned} &-\frac{g_2^4}{2048\pi^2 M^2_{\tilde{W}^-}}\cdot\left(I(y_{\tilde{\nu}_{\mu 1}}, y_{\tilde{\nu}_{e1}}) - I(y_{\tilde{\nu}_{\mu 1}}, y_{\tilde{\nu}_{e3}}) - I(y_{\tilde{\nu}_{\mu 3}}, y_{\tilde{\nu}_{e1}}) + I(y_{\tilde{\nu}_{\mu 3}}, y_{\tilde{\nu}_{e3}})\right)\\ &\approx -\frac{g_2^4}{2048\pi^2 M^2_{\tilde{W}^-}}\cdot(y_{\tilde{\nu}_{\mu 1}} - y_{\tilde{\nu}_{\mu 3}})(y_{\tilde{\nu}_{e1}} - y_{\tilde{\nu}_{e3}})\frac{\partial}{\partial y_{\tilde{\nu}_{\mu 1}}}\frac{\partial}{\partial y_{\tilde{\nu}_{e1}}}I(y_{\tilde{\nu}_{\mu 1}}, y_{\tilde{\nu}_{e1}})\\ &\approx -\frac{g_2^4}{2048\pi^2 M^2_{\tilde{W}^-}}\cdot\frac{\Delta m^2_{\tilde{\nu}_\mu}}{M^2_{\tilde{W}^-}}\cdot\frac{\Delta m^2_{\tilde{\nu}_e}}{M^2_{\tilde{W}^-}}\cdot\frac{\partial}{\partial y_{\tilde{\nu}_{\mu 1}}}\frac{\partial}{\partial y_{\tilde{\nu}_{e1}}}I(y_{\tilde{\nu}_{\mu 1}}, y_{\tilde{\nu}_{e1}}), \end{aligned} \tag{4.56}$$

where the squared-mass differences are

$$\begin{aligned} \Delta m^2_{\tilde{\nu}_\mu} &= \frac{4(m^{\mu}_D)^2(A_\mu - \mu\cot\beta - B_\mu)}{M^{\mu}_R},\\ \Delta m^2_{\tilde{\nu}_e} &= \frac{4(m^{e}_D)^2(A_e - \mu\cot\beta - B_e)}{M^{e}_R}. \end{aligned} \tag{4.57}$$

Assuming $A_\mu = A_e \equiv A$ and $B_\mu = B_e \equiv B$, the squared-mass differences of light muon sneutrinos and light electron sneutrinos are

$$\Delta m^2_{\tilde{\nu}_\mu} = \Delta m^2_{\tilde{\nu}_e} \equiv \Delta m^2_{\tilde{\nu}} = \frac{4m_D^2(A - \mu\cot\beta - B)}{M_R} \tag{4.58}$$

so that Eq. 4.56 then simplifies to

$$-\frac{g_2^4}{2048\pi^2 M_{\tilde{W}^-}^2}\cdot\frac{\Delta m_{\tilde{\nu}_\mu}^2}{M_{\tilde{W}^-}^2}\cdot\frac{\Delta m_{\tilde{\nu}_e}^2}{M_{\tilde{W}^-}^2}\cdot\frac{\partial}{\partial y_{\tilde{\nu}_{\mu 1}}}\frac{\partial}{\partial y_{\tilde{\nu}_{e1}}}I(y_{\tilde{\nu}_{\mu 1}},y_{\tilde{\nu}_{e1}})$$
$$\approx -\frac{g_2^4(\Delta m_{\tilde{\nu}}^2)^2}{2048\pi^2 M_{\tilde{W}^-}^6}\frac{\partial}{\partial y_{\tilde{\nu}_{\mu 1}}}\frac{\partial}{\partial y_{\tilde{\nu}_{e1}}}I(y_{\tilde{\nu}_{\mu 1}},y_{\tilde{\nu}_{e1}}). \quad (4.59)$$

The contribution from graph (e) and (f) is not dominated by the terms involving only light sneutrinos even though $I(z_{\tilde{\nu}_{\mu a}},z_{\tilde{\nu}_{eb}})$ is a decreasing function of $z_{\tilde{\nu}_{\mu a}}$ and $z_{\tilde{\nu}_{eb}}$, because these terms get suppressed by the mixing matrix. The terms including only light sneutrinos $\tilde{\nu}_{\mu 1}$, $\tilde{\nu}_{\mu 3}$, $\tilde{\nu}_{e1}$, $\tilde{\nu}_{e3}$ roughly gives

$$(U_{21}^\mu)^2(U_{21}^e)^2 I(z_{\tilde{\nu}_{\mu 1}},z_{\tilde{\nu}_{e1}})-(U_{21}^\mu)^2(U_{43}^e)^2 I(z_{\tilde{\nu}_{\mu 1}},z_{\tilde{\nu}_{e3}})$$
$$-(U_{43}^\mu)^2(U_{21}^e)^2 I(z_{\tilde{\nu}_{\mu 3}},z_{\tilde{\nu}_{e1}})+(U_{43}^\mu)^2(U_{43}^e)^2 I(z_{\tilde{\nu}_{\mu 3}},z_{\tilde{\nu}_{e3}})$$
$$\approx\frac{m_D^4}{M_R^4}\cdot\frac{\Delta m_{\tilde{\nu}_\mu}^2\Delta m_{\tilde{\nu}_e}^2}{M_{\tilde{h}_B^-}^4}\frac{\partial}{\partial z_{\tilde{\nu}_{\mu 1}}}\frac{\partial}{\partial z_{\tilde{\nu}_{e1}}}I(z_{\tilde{\nu}_{\mu 1}},z_{\tilde{\nu}_{e1}})$$
$$\sim\mathcal{O}\left(\frac{1}{M_R^6}\right), \quad (4.60)$$

while the terms including one light and one heavy sneutrino are roughly

$$(U_{21}^\mu)^2(U_{22}^e)^2 I(z_{\tilde{\nu}_{\mu 1}},z_{\tilde{\nu}_{e2}})-(U_{21}^\mu)^2(U_{44}^e)^2 I(z_{\tilde{\nu}_{\mu 1}},z_{\tilde{\nu}_{e4}})$$
$$-(U_{43}^\mu)^2(U_{22}^e)^2 I(z_{\tilde{\nu}_{\mu 3}},z_{\tilde{\nu}_{e2}})+(U_{43}^\mu)^2(U_{44}^e)^2 I(z_{\tilde{\nu}_{\mu 3}},z_{\tilde{\nu}_{e4}})$$
$$+(U_{22}^\mu)^2(U_{21}^e)^2 I(z_{\tilde{\nu}_{\mu 2}},z_{\tilde{\nu}_{e1}})-(U_{22}^\mu)^2(U_{43}^e)^2 I(z_{\tilde{\nu}_{\mu 2}},z_{\tilde{\nu}_{e3}})$$
$$-(U_{44}^\mu)^2(U_{21}^e)^2 I(z_{\tilde{\nu}_{\mu 4}},z_{\tilde{\nu}_{e1}})+(U_{44}^\mu)^2(U_{43}^e)^2 I(z_{\tilde{\nu}_{\mu 4}},z_{\tilde{\nu}_{e3}})$$
$$\approx\left(\frac{m_D^e}{M_R^e}\right)^2\frac{\Delta M_{\tilde{\nu}_\mu}^2\cdot\Delta m_{\tilde{\nu}_e}^2}{M_{\bar{B}}^4}\cdot\left(\frac{M_{\bar{B}}}{M_R^\mu}\right)^4+\left(\frac{m_D^\mu}{M_R^\mu}\right)^2\frac{\Delta M_{\tilde{\nu}_e}^2\cdot\Delta m_{\tilde{\nu}_\mu}^2}{M_{\bar{B}}^4}\cdot\left(\frac{M_{\tilde{h}_B^-}}{M_R^e}\right)^4$$
$$\sim\mathcal{O}\left(\frac{1}{M_R^6}\right), \quad (4.61)$$

where $\Delta M_{\tilde{\nu}_e}^2$ and $\Delta M_{\tilde{\nu}_e}^2$ are the heavy sneutrino squared-mass differences

$$\Delta M_{\tilde{\nu}_e}^2=4B_eM_R^e \quad\text{and}\quad \Delta M_{\tilde{\nu}_\mu}^2=4B_\mu M_R^\mu. \quad (4.62)$$

Under our approximations,

$$\Delta M_{\tilde{\nu}_e}^2=\Delta M_{\tilde{\nu}_\mu}^2\equiv\Delta M_{\tilde{\nu}}^2=4BM_R. \quad (4.63)$$

The terms including two heavy sneutrinos are roughly

$$\begin{aligned}
&(U^{\mu}_{22})^2(U^{e}_{22})^2 I(z_{\tilde{\nu}_{\mu 2}}, z_{\tilde{\nu}_{e2}}) - (U^{\mu}_{22})^2(U^{e}_{44})^2 I(z_{\tilde{\nu}_{\mu 2}}, z_{\tilde{\nu}_{e4}}) \\
&- (U^{\mu}_{44})^2(U^{e}_{22})^2 I(z_{\tilde{\nu}_{\mu 4}}, z_{\tilde{\nu}_{e2}}) + (U^{\mu}_{44})^2(U^{e}_{44})^2 I(z_{\tilde{\nu}_{\mu 4}}, z_{\tilde{\nu}_{e4}}) \\
&\approx \frac{\Delta M^2_{\tilde{\nu}_{\mu}} \cdot \Delta M^2_{\tilde{\nu}_{e}}}{M^4_{\tilde{B}}} \cdot \frac{M^6_{\tilde{h}_{\bar{B}}}}{3M^6_R} \\
&\approx \frac{(\Delta M^2_{\tilde{\nu}})^2 M^2_{\tilde{B}}}{3M^6_R} \\
&\sim \mathcal{O}\left(\frac{1}{M^4_R}\right).
\end{aligned} \tag{4.64}$$

Comparing the M_R dependences of Eqs. 4.60, 4.61 and 4.64, we see that the dominant term is the one involving two heavy sneutrinos. Thus the contribution from graph (e) and (f) can be approximated as

$$\begin{aligned}
&-\frac{(m^{\mu}_D)^2(m^{e}_D)^2}{2048 V_T^4 \pi^2 M^2_{\tilde{h}_{\bar{B}}}} \cdot \left(\sum_{a=1}^{2}\sum_{b=1}^{2} (U^{\mu}_{2a})^2(U^{e}_{2b})^2 I(z_{\tilde{\nu}_{\mu a}}, z_{\tilde{\nu}_{eb}}) - \sum_{a=1}^{2}\sum_{b=3}^{4} (U^{\mu}_{2a})^2(U^{e}_{4b})^2 I(z_{\tilde{\nu}_{\mu a}}, z_{\tilde{\nu}_{eb}}) \right. \\
&\left. - \sum_{a=3}^{4}\sum_{b=1}^{2} (U^{\mu}_{4a})^2(U^{e}_{2b})^2 I(z_{\tilde{\nu}_{\mu a}}, z_{\tilde{\nu}_{eb}}) + \sum_{a=3}^{4}\sum_{b=3}^{4} (U^{\mu}_{4a})^2(U^{e}_{4b})^2 I(z_{\tilde{\nu}_{\mu a}}, z_{\tilde{\nu}_{eb}}) \right) \\
&\approx -\frac{m^4_D (\Delta M^2_{\tilde{\nu}})^2}{6144 V_T^4 \pi^2 M^6_R} = -\frac{g_2^4 m^4_D (\Delta M^2_{\tilde{\nu}})^2 (1+\tan^2\beta)^2}{24576\pi^2 M^6_R M^4_W \tan^4\beta}.
\end{aligned} \tag{4.65}$$

Combining the various contributions, the effective coupling constant is thus roughly given by

$$\begin{aligned}
|ReG_{\bar{M}M}| \approx &\left| \frac{g_2^4 m^4_D}{1024\pi^2 M^2_R M^4_W} \cdot \ln\frac{M_R}{M_W} - \frac{g_2^4 (\Delta m^2_{\tilde{\nu}})^2}{2048\pi^2 M^6_{\tilde{W}^-}} \cdot \frac{\partial}{\partial y_{\tilde{\nu}_{\mu 1}}}\frac{\partial}{\partial y_{\tilde{\nu}_{e1}}} I(y_{\tilde{\nu}_{\mu 1}}, y_{\tilde{\nu}_{e1}}) \right. \\
&\left. - \frac{g_2^4 m^4_D (\Delta M^2_{\tilde{\nu}})^2 (1+\tan^2\beta)^2}{24576\pi^2 M^6_R M^4_W \tan^4\beta} \right|.
\end{aligned} \tag{4.66}$$

The first term in Eq. 4.66 is the dominant contribution of graph (a) and (b), which contains intermediate neutrino and W boson. This contribution appears in the model in which all SUSY partners decoupled. The second term is the dominant contribution of graph (c) and (d), in which wino and sneutrino appear in the intermediate states. Finally, the third term is the dominant contribution of graph (e) and (f), with intermediate higgsino and sneutrinos lines. The second and third terms both depend on the sneutrino mass splitting. This reflects the intra-generation lepton number violating property of the muonium–Antimuonium oscillation process, because the sneutrino mass splitting is generated by the $\Delta L = 2$ operators in the sneutrino mass matrix. To compare the relative sizes of these three terms, we use the current experimental limits of the neutrino and sparticle masses.

The first terms in Eq. 4.66 has a factor m_D^4/M_R^2, which is the scale of the light neutrino mass square $m_\nu^2 \simeq m_D^4/M_R^2$ generated by see–saw mechanism. The experimental constraints on neutrino masses are summarized in reference (Vogel and Piepke 2008) as

$$\begin{aligned} &m_\nu(\text{electron based}) < 225\,\text{eV}, \\ &m_\nu(\text{muon based}) < 0.19\,\text{MeV}, \\ &m_\nu(\text{tau based}) < 18.2\,\text{MeV}, \end{aligned} \tag{4.67}$$

For instance, assuming

$$m_D \sim M_W, \tag{4.68}$$

$$m_\nu = \frac{m_D^2}{M_R} \sim 1\,\text{eV}, \tag{4.69}$$

then the right-handed neutrino mass scale is about

$$M_R \sim 10^{13}\,\text{GeV}. \tag{4.70}$$

In this case, the first terms in Eq. 4.66 is roughly

$$\frac{g_2^4 m_D^4}{1024\pi^2 M_R^2 M_W^4} \cdot \ln\frac{M_R}{M_W} = \frac{g_2^4 m_\nu^2}{1024\pi^2 M_W^4} \cdot \ln\frac{M_R}{M_W} \simeq 1.1 \times 10^{-29}\,\text{GeV}^{-2}. \tag{4.71}$$

The second term in Eq. 4.66 depends on the light sneutrino squared-mass difference $\Delta m_{\tilde\nu}^2$, which can be written in terms of light sneutrino mass splitting $\Delta m_{\tilde\nu}$ by

$$\Delta m_{\tilde\nu}^2 = 2 m_{\tilde\nu} \Delta m_{\tilde\nu}, \tag{4.72}$$

where $m_{\tilde\nu}$ is the mass scale of light sneutrinos. So doing the second term in Eq. 4.66 can be written as

$$-\frac{g_2^4 (\Delta m_{\tilde\nu}^2)^2}{2048\pi^2 M_{\tilde W^-}^6} \cdot \frac{\partial}{\partial y_{\tilde\nu_{\mu 1}}} \frac{\partial}{\partial y_{\tilde\nu_{e1}}} I(y_{\tilde\nu_{\mu 1}}, y_{\tilde\nu_{e1}}) = -\frac{g_2^4 m_{\tilde\nu}^2 (\Delta m_{\tilde\nu})^2}{512\pi^2 M_{\tilde W^-}^6} \cdot \frac{\partial}{\partial y_{\tilde\nu_{\mu 1}}} \frac{\partial}{\partial y_{\tilde\nu_{e1}}} I(y_{\tilde\nu_{\mu 1}}, y_{\tilde\nu_{e1}}). \tag{4.73}$$

Yuval and Howard (1997) provide an upper limit on the sneutrino mass splitting by calculating the one-loop correction to the neutrino mass. Assuming that this correction is no larger than the tree result gives

$$\Delta m_{\tilde\nu} \le 2 \times 10^3 m_\nu. \tag{4.74}$$

Relaxing this absence of fine tuning constraint can substantially enhance the contribution of the graph (c) and (d). Taking the sneutrino mass splitting to be of the same order as sneutrino mass

$$\Delta m_{\tilde{\nu}} \sim m_{\tilde{\nu}}, \tag{4.75}$$

and $m_{\tilde{\nu}_\mu}, m_{\tilde{\nu}_e}$ to be the common mass scale $m_{\tilde{\nu}}$ gives

$$y_{\tilde{\nu}_{\mu 1}} \sim y_{\tilde{\nu}_{e1}} \sim y_{\tilde{\nu}} = \frac{m_{\tilde{\nu}}^2}{M_{\tilde{W}^-}^2}. \tag{4.76}$$

Equation 4.73 can then be written as

$$-\frac{g_2^4 m_{\tilde{\nu}}^2 (\Delta m_{\tilde{\nu}})^2}{512\pi^2 M_{\tilde{W}^-}^6} \cdot \frac{\partial}{\partial y_{\tilde{\nu}_{\mu 1}}} \frac{\partial}{\partial y_{\tilde{\nu}_{e1}}} I(y_{\tilde{\nu}_{\mu 1}}, y_{\tilde{\nu}_{e1}}) \approx -\frac{g_2^4 m_{\tilde{\nu}}^4}{512\pi^2 M_{\tilde{W}^-}^6} \cdot \frac{\partial}{\partial y_{\tilde{\nu}_{\mu 1}}} \frac{\partial}{\partial y_{\tilde{\nu}_{e1}}} I(y_{\tilde{\nu}_{\mu 1}}, y_{\tilde{\nu}_{e1}})\Big|_{y_{\tilde{\nu}_{\mu 1}}, y_{\tilde{\nu}_{e1}} = y_{\tilde{\nu}}}. \tag{4.77}$$

The function $\frac{m_{\tilde{\nu}}^4}{M_{\tilde{W}^-}^6} \cdot \frac{\partial}{\partial y_{\tilde{\nu}_{\mu 1}}} \frac{\partial}{\partial y_{\tilde{\nu}_{e1}}} I(y_{\tilde{\nu}_{\mu 1}}, y_{\tilde{\nu}_{e1}})\Big|_{y_{\tilde{\nu}_{\mu 1}}, y_{\tilde{\nu}_{e1}} = y_{\tilde{\nu}}}$ in Eq. 4.77 is a decreasing function of $M_{\tilde{W}^-}$. In order to calculate the maximum contribution of graph (c) and (d), we use the experimental lower bound on $M_{\tilde{W}^-}$. Many experimental searches for physics beyond the standard model have been conducted and provide various constraints on SUSY parameter space. Table 4.1 lists some of the constraints (Grivaz 2008). Fixing the wino mass to its lower limit in Table 4.1.

$$M_{\tilde{W}^-} = 94.0\,\text{GeV}, \tag{4.78}$$

the contribution $\frac{g_2^4 m_{\tilde{\nu}}^4}{512\pi^2 M_{\tilde{W}^-}^6} \cdot \frac{\partial}{\partial y_{\tilde{\nu}_{\mu 1}}} \frac{\partial}{\partial y_{\tilde{\nu}_{e1}}} I(y_{\tilde{\nu}_{\mu 1}}, y_{\tilde{\nu}_{e1}})\Big|_{y_{\tilde{\nu}_{\mu 1}}, y_{\tilde{\nu}_{e1}} = y_{\tilde{\nu}}}$ as a function of sneutrino mass scale $m_{\tilde{\nu}}$ is shown in Fig. 4.2. When $m_{\tilde{\nu}} = 94.0\,\text{GeV}$, which is allowed by the experimental limit in Table 4.1, the contribution of graph (c) and (d) reaches its maximum so that

$$\frac{g_2^4 m_{\tilde{\nu}}^4}{512\pi^2 M_{\tilde{W}^-}^6} \cdot \frac{\partial}{\partial y_{\tilde{\nu}_{\mu 1}}} \frac{\partial}{\partial y_{\tilde{\nu}_{e1}}} I(y_{\tilde{\nu}_{\mu 1}}, y_{\tilde{\nu}_{e1}})\Big|_{y_{\tilde{\nu}_{\mu 1}}, y_{\tilde{\nu}_{e1}} = y_{\tilde{\nu}}} \leq 1.3 \times 10^{-10}\,\text{GeV}^{-2}. \tag{4.79}$$

Table 4.1 Experimental lower limits on SUSY particle masses

Sparticle	Lower limit (GeV)
$\tilde{\chi}_1^{\pm}$	94.0
$\tilde{\chi}_1^0$	46.0
$\tilde{\nu}$	94.0
$\tilde{\mu}_R$	94.0
$\tilde{e}_L$	107.0
$\tilde{e}_R$	73.0

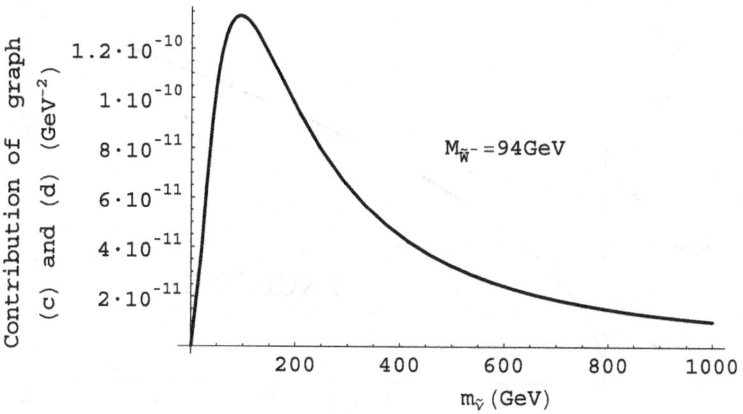

Fig. 4.2 The contribution of graph (c) and (d) as a function of $m_{\tilde{\nu}}$ when fixing the wino mass to its lower limit $M_{\tilde{W}^-}$

Finally, the third term in Eq. 4.66 depends on the heavy sneutrino squared-mass difference $\Delta M_{\tilde{\nu}}^2 = 2M_{\tilde{\nu}}\Delta M_{\tilde{\nu}} = 4BM_R$. Since we assume that M_R is the largest mass scale, $\Delta M_{\tilde{\nu}}$ can't be arbitrarily large. Taking parameter B one order of magnitude smaller than M_R, the heavy sneutrino mass splitting is

$$\Delta M_{\tilde{\nu}} \sim \frac{M_R}{10}. \tag{4.80}$$

The contribution of graph (e) and (f) can then be written as

$$\frac{g_2^4 m_D^4 (\Delta M_{\tilde{\nu}}^2)^2 (1+\tan^2\beta)^2}{24576\pi^2 M_R^6 M_W^4 \tan^4\beta} = \frac{g_2^4 m_\nu^2 (1+\tan^2\beta)^2}{614400\pi^2 M_W^4 \tan^4\beta}. \tag{4.81}$$

When $\tan\beta$ is very small the contribution can get large and even reach the experimental limit Eq. 4.51. In this case, the experimental limit of muonium–antimuonium oscillation provides an inequality relating $\tan\beta$ and m_ν, which is given by

$$\frac{g_2^4 m_\nu^2 (1+\tan^2\beta)^2}{614400\pi^2 M_W^4 \tan^4\beta} \leq 3.5 \times 10^{-8}\,\text{GeV}^{-2}. \tag{4.82}$$

This inequality translates into a lower bound of $\tan\beta$ for different light neutrino masses m_ν:

$$\begin{aligned} \tan\beta &\geq 3.7\times10^{-7}, && \text{if } m_\nu = 1\,\text{eV},\\ \tan\beta &\geq 3.7\times10^{-8}, && \text{if } m_\nu = 10^{-2}\,\text{eV},\\ \tan\beta &\geq 3.7\times10^{-9}, && \text{if } m_\nu = 10^{-4}\,\text{eV}. \end{aligned} \tag{4.83}$$

The lower limit on $\tan\beta$ as a function of light neutrino mass scale m_ν is shown in Fig. 4.3. Notice that the ratio of the two Higgs VEVs $\tan\beta$ are related to the light

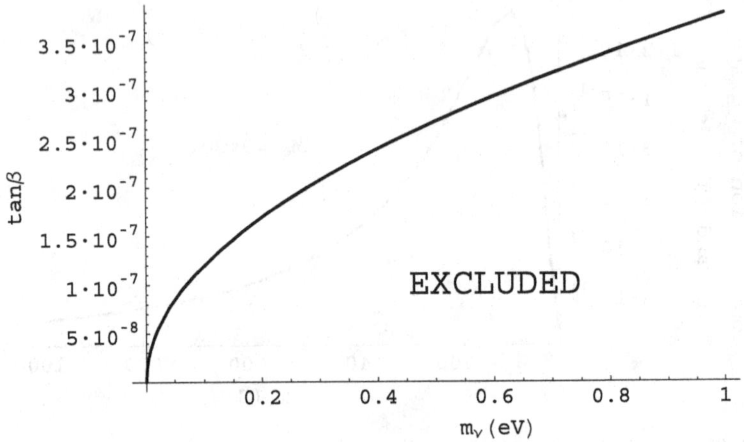

Fig. 4.3 The lower limit on $\tan\beta$ as a function of light neutrino mass scale m_ν provided by the muonium–antimuonium oscillation experiment. The area above the curve is allowed by the experiment results

neutrino masses in the above inequality, although the graph (e) and (f) do not involve any neutrinos in the intermediate states. This results since we are using a specific model where the neutrino masses are generated by see–saw mechanism $m_\nu \sim \mathcal{O}(m_D^2/M_R)$. The sneutrino mixing matrix is approximated in term of m_D/M_R and the heavy sneutrino masses are also of order M_R. If we take m_D to be of order M_W, the heavy sneutrino masses M_R in the contribution of graph (e) and (f) can be expressed in term of the light neutrino mass scale m_ν. This explains the appearance of the parameter m_ν in the inequality Eq. 4.82.

However, for non-infinitesimal values of $\tan\beta$, this contribution is very small compared with the maximum of the second term in Eq. 4.66. For instance, taking the neutrino mass m_ν to be 1 eV and assuming $\tan\beta \geq 10^{-4}$, the contribution of graph (e) and (f) is

$$\frac{g_2^4 m_\nu^2 (1+\tan^2\beta)^2}{614400\pi^2 M_W^4 \tan^4\beta} \lesssim 7.2 \times 10^{-18}\,\text{GeV}^{-2}. \tag{4.84}$$

Thus, except for the case of very small $\tan\beta$, the second term in Eq. 4.66 is the dominant contribution for a wide range of the parameters and its maximum is roughly two orders of magnitude below the sensitivity of the current experiments.

References

P. Minkowski, Phys. Lett. **B67**, 421 (1977)

T. Yanagida, In: Proceedings of the Workshop on the Unified Theories and Baryon Number in the Universe, Tsukuba, Japan, 1979, (O. Sawada and A. Sugamoto eds), p 95

M. Gell-Mann, P. Ramond and R. Slansky in *Supergravity*, (P. van Nieuwenhuizen and D. Freedman eds) North-Holland, Amsterdam, 1979
S.L. Glashow, In: Proceedings of the 1979 Cargese Institute on Quarks and Leptons (M. Levy et. al. eds.) Plenum Press, New York 1980, p 687
R.N. Mohapatra and G. Senjanovic. Phys. Rev. Lett. **44**, 912 (1980).
Y. Grossman and HE.Haber, Phys.Rev. Lett. **78**, 3438 (1997).
L. Willmann et al, Phys. Rev. Lett **82**, 49 (1999)
P. Vogel and A. Piepke, Phys. Lett. **B667**, 517 (2008)
J.-F.Grivaz, Phys. Lett. **B667**, 1228 (2008)

Chapter 5
The Constraints from the Muon and Electron Anomalous Magnetic Moment Experiments

One has to be careful about other constraints on the model parameters. Examples of such potential constraints come from the measurements of the muon and electron anomalous magnetic moments. The correction to the muon anomalous magnetic moment in the model under consideration is found by calculating the one-loop graphs shown in Fig. 5.1.

The muon anomalous magnetic moment contributed from the above graphs is

$$\begin{aligned} a_\mu^{BSM} &= -\frac{g_2^2 m_\mu^2}{16\pi^2 M_W^2}(V_{12}^\mu V_{12}^{\mu *})^2 F^W(x_{\nu_{\mu 2}}) \\ &+ \frac{g_2^2 m_\mu^2}{32\pi^2 M_{\tilde{W}^-}^2}\left(\sum_{a=1}^{2}(U_{1a}^\mu)^2 F^C(y_{\tilde{\nu}_{\mu a}}) + \sum_{a=3}^{4}(U_{3a}^\mu)^2 F^C(y_{\tilde{\nu}_{\mu a}})\right) \\ &+ \frac{(m_D^\mu)^2 m_\mu^2}{32\pi^2 V_T M_{\tilde{h}_{\bar{B}}^-}^2}\left(\sum_{a=1}^{2}(U_{2a}^\mu)^2 F^C(z_{\tilde{\nu}_{\mu a}}) + \sum_{a=3}^{4}(U_{4a}^\mu)^2 F^C(z_{\tilde{\nu}_{\mu a}})\right) \\ &- \frac{g_2^2 m_\mu^2}{32\pi^2 M_{\tilde{W}^0}^2}F^N(s_{\tilde{\mu}_L}) - \frac{g_2^2 m_\mu^2}{32\pi^2 M_{\tilde{B}}^2}F^N(t_{\tilde{\mu}_L}) - \frac{g_1^2 m_\mu^2}{8\pi^2 M_{\tilde{B}}^2}F^N(t_{\tilde{\mu}_R}), \end{aligned} \tag{5.1}$$

where $s_{\tilde{\mu}_a} = \frac{m_{\tilde{\mu}_a}^2}{M_{\tilde{W}^0}^2}$, $t_{\tilde{\mu}_a} = \frac{m_{\tilde{\mu}_a}^2}{M_{\tilde{B}}^2}$, and

$$F^W(x_{\nu_{\mu 2}}) = \int_0^1 dx \frac{-4x^2(1+x) - 2x_\mu \cdot x^2(x-1) - 2x_{\nu_{\mu 2}}(2x - 3x^2 + x^3)}{x_\mu \cdot x^2 + (1 - x_\mu)x + x_{\nu_{\mu 2}}(1-x)}, \tag{5.2}$$

$$F^C(y_{\tilde{\nu}_{\mu a}}) = \frac{2y_{\tilde{\nu}_{\mu a}}^3 - 3y_{\tilde{\nu}_{\mu a}}^2(-1 + 2\ln y_{\tilde{\nu}_{\mu a}}) - 6y_{\tilde{\nu}_{\mu a}} + 1}{6(1 - y_{\tilde{\nu}_{\mu a}})^4}, \tag{5.3}$$

B. Liu, *Muonium–Antimuonium Oscillations in an Extended Minimal Supersymmetric Standard Model*, Springer Theses,
DOI: 10.1007/978-1-4419-8330-5_5, © Springer Science+Business Media, LLC 2011

$$F^N(s_{\tilde{\mu}_a}) = \frac{s_{\tilde{\mu}_a}^3 - 6s_{\tilde{\mu}_a}^2 + 3s_{\tilde{\mu}_a} + 6s_{\tilde{\mu}_a}\ln s_{\tilde{\mu}_a} + 2}{6(1 - s_{\tilde{\mu}_a})^4}. \tag{5.4}$$

With assumption that M_R is the largest mass scale, the dominant contribution of the graphs in Fig. 5.1 to a_μ^{BSM} is

$$\begin{aligned} a_\mu^{BSM} \approx\ & \frac{g_2^2 m_\mu^2 m_D^2}{12\pi^2 M_W^2 M_R^2} + \frac{g_2^2 m_\mu^2}{32\pi^2 M_{\tilde{W}^-}^2} \cdot \left(F^C(y_{\tilde{\nu}_{\mu 1}}) + F^C(y_{\tilde{\nu}_{\mu 3}})\right) + \frac{g_2^2 m_\mu^2 m_D^2 (1 + \tan^2\beta)}{96\pi^2 M_W^2 M_R^2 \tan^2\beta} \\ & - \frac{g_2^2 m_\mu^2}{32\pi^2 M_{\tilde{W}^0}^2} F^N(s_{\tilde{\mu}_L}) - \frac{g_2^2 m_\mu^2}{32\pi^2 M_{\tilde{B}}^2} F^N(t_{\tilde{\mu}_L}) - \frac{g_1^2 m_\mu^2}{8\pi^2 M_{\tilde{B}}^2} F^N(t_{\tilde{\mu}_R}). \end{aligned} \tag{5.5}$$

The second and the last three terms are all decreasing functions of slepton, chargino and neutralino masses. We can use the experimental bounds in Table 4.1 to calculate the maximum values of these terms yielding

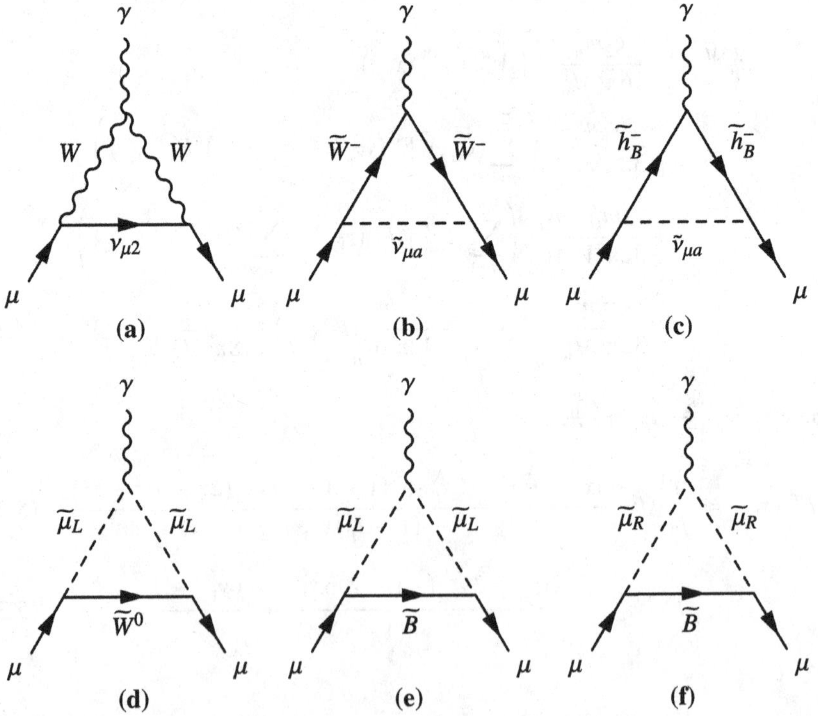

Fig. 5.1 The Feynman graphs contributing to the muon anomalous magnetic moment beyond the Standard Model

$$\frac{g_2^2 m_\mu^2}{32\pi^2 M_{\tilde{W}^-}^2} \cdot \left(F^C(y_{\tilde{\nu}_{\mu 1}}) + F^C(y_{\tilde{\nu}_{\mu 3}})\right) \leq 2.5 \times 10^{-10},$$
$$\frac{g_2^2 m_\mu^2}{32\pi^2 M_{\tilde{W}^0}^2} F^N(s_{\tilde{\mu}_L}) \leq 1.9 \times 10^{-10},$$
$$\frac{g_2^2 m_\mu^2}{32\pi^2 M_{\tilde{B}}^2} F^N(t_{\tilde{\mu}_L}) \leq 1.9 \times 10^{-10},$$
$$\frac{g_1^2 m_\mu^2}{8\pi^2 M_{\tilde{B}}^2} F^N(t_{\tilde{\mu}_R}) \leq 2.1 \times 10^{-10}. \tag{5.6}$$

The maxima of these terms are all about one order of magnitude smaller than the present experimental bound on the contribution to $a_\mu = \frac{1}{2}(g-2)$ beyond the Standard Model (Melnikov and Arkady 2006):

$$\delta a_\mu = a_\mu^{\exp} - a_\mu^{SM} = 2 \times 10^{-9}. \tag{5.7}$$

The first and third term both depend on the light neutrino mass scale m_ν and get suppressed. For instance, using the assumptions Eqs. (4.68) and (4.69), the first and third terms are

$$\frac{g_2^2 m_\mu^2 m_D^2}{12\pi^2 M_W^2 M_R^2} = \frac{g_2^2 m_\mu^2 m_\nu^2}{12\pi^2 M_W^4} \approx 8.7 \times 10^{-31},$$
$$\frac{g_2^2 m_\mu^2 m_D^2 (1+\tan^2\beta)}{96\pi^2 M_W^2 M_R^2 \tan^2\beta} = \frac{g_2^2 m_\mu^2 m_\nu^2 (1+\tan^2\beta)}{96\pi^2 M_W^4 \tan^2\beta} \approx 1.1 \times 10^{-31} \cdot \frac{1+\tan^2\beta}{\tan^2\beta}. \tag{5.8}$$

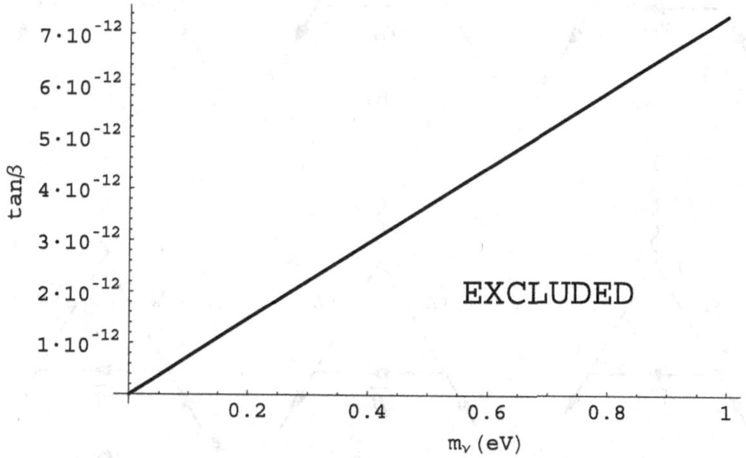

Fig. 5.2 The lower limit on $\tan\beta$ as a function of light neutrino mass scale m_ν provided by the muon anomalous magnetic moment experiment. The area above the curve is allowed by the experiment results

The first term is negligible compared with the terms in Eq. (5.7). However, the third term can be large if $\tan\beta$ is very small. Therefore, the experimental bound on the muon magnetic moment will provide an inequality on $\tan\beta$ and m_ν, which is given by

$$\frac{g_2^2 m_\mu^2 m_\nu^2 (1+\tan^2\beta)}{96\pi^2 M_W^4 \tan^2\beta} \leq 2\times 10^{-9}. \tag{5.9}$$

This inequality translates into a lower bound on $\tan\beta$ as a function of the light neutrino mass scale m_ν as shown in Fig. 5.2.

The electron anomalous magnetic moment beyond the Standard Model is contributed by the six Feynman graphs displayed in Fig. 5.3. The experimental bound on the contribution to the electron anomalous magnetic moment beyond Standard Model is (Ellis et al. 1994)

$$\delta a_e = a_e^{\exp} - a_e^{SM} = 1.4\times 10^{-11}. \tag{5.10}$$

In analogy to the muon case, this experimental limit will also generate an inequality relation of m_ν and $\tan\beta$ given by

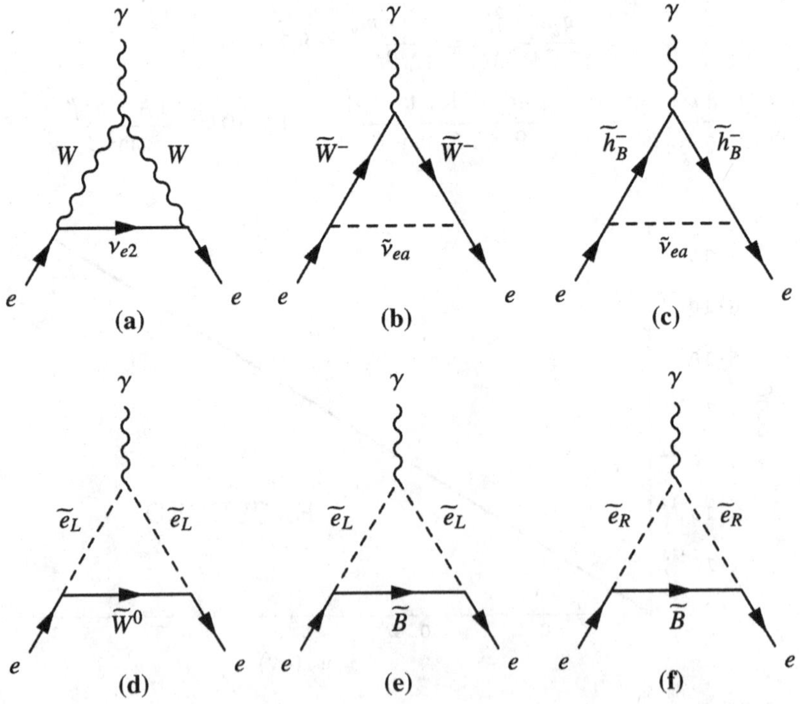

Fig. 5.3 The Feynman graphs contributing to the electron anomalous magnetic moment beyond the Standard Model

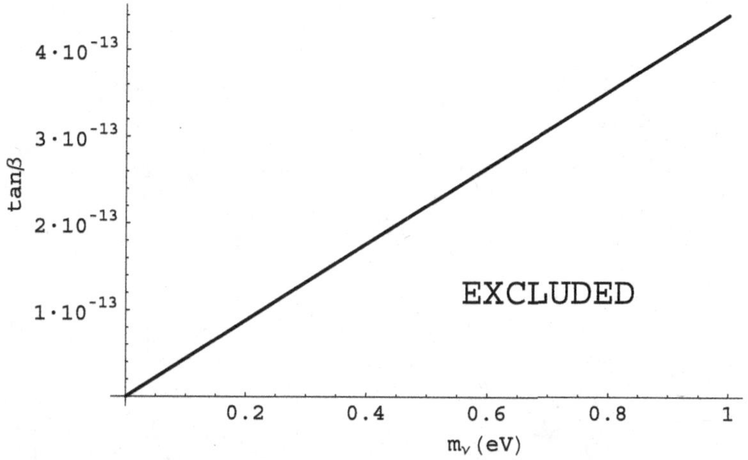

Fig. 5.4 The lower limit on $\tan\beta$ as a function of light neutrino mass scale m_ν provided by the electron anomalous magnetic moment experiment. The area above the curve is allowed by the experiment results

$$\frac{g_2^2 m_e^2 m_\nu^2 (1 + \tan^2\beta)}{96\pi^2 M_W^4 \tan^2\beta} \leq 1.4 \times 10^{-11}. \tag{5.11}$$

This inequality is illustrated in Fig. 5.4.

From Figs. 4.3, 5.2 and 5.4 we see that for the model we are considering, the muonium–antimuonium oscillation experiment gives a more stringent constraint on $\tan\beta$ than the muon and electron anomalous magnetic moment experiments.

References

J.R. Ellis, M. Karliner, M.A. Samuel and E. Steinfelds, SLAC-PUB-6670, CERN-TH-7451-94, TAUP-2001-94, OSU-RN-293, hep-ph/9409376, 1994.

K. Melnikov and A. Vainshtein, *Theory of the Muon Anomalous Magnetic Moment*, (Springer), 2006, p. 152.

Chapter 6
Conclusions

I have calculated the effective coupling constant of the muonium–antimuonium oscillation process in two different models. First, I modified the Standard Model by including three singlet right-handed neutrinos. Present experimental limits resulting from the non-observation of the oscillation process sets a lower limit on the right-handed neutrino mass scale M_R roughtly of order 1 TeV. Second, the muonium–antimuonium oscillation was investigated in the Minimal Supersymmetric Standard Model extended by inclusion of three right-handed neutrino superfields where the required lepton flavor violation has its origin in the Majorana property of the neutrino and sneutrino mass eigenstates. For a wide range of the parameters, the contribution of the graphs mediated by the sneutrino and winos $\widetilde{W}^-$ is dominant. The maximum of this contribution to the effective coupling constant is roughly two orders of magnitude below the sensitivity of current muonium–antimuonium oscillation experiments. However, there is very limited possibility that the contribution of the graphs mediated by sneutrinos and Higgsino $\widetilde{h}_B^-$ is dominant if $\tan\beta$ is very small. In this case, the contributions can even be large enough to reach the present experimental bound. Therefore, the experimental bound can provide an inequality on the model parameters, which can be translated into a lower bound on $\tan\beta$ as a function of light neutrino mass scale m_ν. The constraints from the muon and electron anomalous magnetic moments were also investigated. For this model, the muonium–antimuonium oscillation experiments give the most stringent constraints on the parameters.

B. Liu, *Muonium–Antimuonium Oscillations in an Extended Minimal Supersymmetric Standard Model*, Springer Theses,
DOI: 10.1007/978-1-4419-8330-5_6,

Appendices

Appendix A. Proof of Identity (2.22)

Using the definition of the mixing matrix $V_{aA} = \sum_{c=1}^{3} (A_L^{-1})_{ac} U_{cA}$, one can write

$$\sum_{A=1}^{6} V_{aA} V_{bA} m = \sum_{A=1}^{6} \left(\sum_{c=1}^{3} (A_L^{-1})_{ac} U_{cA} \right) \cdot \left(\sum_{d=1}^{3} (A_L^{-1})_{bd} U_{dA} \right)_{\nu A}$$

$$= \sum_{c=1}^{3} \sum_{d=1}^{3} (A_L^{-1})_{ac} \left(\sum_{A=1}^{6} U_{cA} m U_{dA} \right) (A_L^{-1})_{bd} \tag{A.1}$$

where $m_{\nu A}$ are the diagonal elements of matrix M^{ν}_{diag},

$$m_{\nu A} = (M^{\nu}_{\text{diag}})_{AA}. \tag{A.2}$$

Consequently, we can express $\sum_{A=1}^{6} U_{cA} m_{\nu A} U_{dA}$ as a product of matrices and Eq. (A.1) takes the form

$$\sum_{A=1}^{6} V_{aA} V_{bA} m_{\nu A} = \sum_{c=1}^{3} \sum_{d=1}^{3} (A_L^{-1})_{ac} \left(U M^{\nu}_{\text{diag}} U^T \right)_{cd} (A_L^{-1})_{bd} \tag{A.3}$$

Using Eq. (2.9), $M^{\nu}_{\text{diag}} = U^T M^{\nu} U$, it follows that

$$M^{\nu *} = U M^{\nu}_{\text{diag}} U^T \tag{A.4}$$

Substituting this result back into Eq. (A.3) then gives

$$\sum_{A=1}^{6} V_{aA} V_{bA} m_{\nu A} = \sum_{c=1}^{3} \sum_{d=1}^{3} (A_L^{-1})_{ac} (M^{\nu *})_{cd} (A_L^{-1})_{bd} \tag{A.5}$$

where

$$M^{\nu *} = \begin{pmatrix} 0 & (m^D)^{T*} \\ m^{D*} & m^{R*} \end{pmatrix} \tag{A.6}$$

Since c and d both run from 1 to 3, $(M^{\nu *})_{cd}$ are the elements of the upper left 3×3 block of matrix (A.6), which is zero. Hence, we secure the identity

$$\sum_{A=1}^{6} V_{aA} V_{bA} m_{\nu A} = 0 \tag{A.7}$$

Appendix B. The Calculations of T-Matrix Elements of the Graphs in Fig. 2.1

B.1 T-Matrix Element of Graph (a1) in Fig. 2.2

The T-matrix element of graph $(a1)$ is:

$$\begin{aligned} iT_{a1} = & \left(-\frac{ig}{\sqrt{2}}\right)^4 \int \frac{d^4p}{(2\pi)^4} \sum_{A=1}^{6} \bar{\mu}_L(3)\gamma_{\mu A} \frac{\gamma p + m_A}{p^2 - m_A^2} V_{eA}^* \gamma_L(2) \\ & \cdot \sum_{B=1}^{6} \bar{\mu}_L(4)\gamma_\rho V \frac{\gamma p + m_B}{p^2 - m_B^2} \gamma_\nu V_{eB}^* e_L(1) \cdot \frac{1}{p^2 - M_W^2} \left[g^{\mu\nu} + \frac{(\xi - 1)p^\nu}{p^2 - \xi M_W^2} \right] \\ & \cdot \frac{1}{p^2 - M_W^2} \left[g^{\rho\sigma} + \frac{(\xi - 1)p^\sigma}{p^2 - \xi M_W^2} \right] \end{aligned} \tag{B.1}$$

We multiply out everything and write above equation in an explicit form:

$$\begin{aligned} iT_{a1} = & \left(\frac{g}{\sqrt{2}}\right)^4 \sum_{A=1}^{6} \sum_{B=1}^{6} (V_{\mu A} V_{eA}^*)(V_{\mu B} V_{eB}^*) \int \frac{d^4p}{(2\pi)^4} \frac{1}{(p^2 - m_A^2)(p^2 - m_B^2)(p^2 - M_W^2)^2} \\ & \cdot \Big\{ [\bar{\mu}(3)\gamma_\mu \gamma p \gamma_\sigma \frac{1-\gamma_5}{2} e(2)][\bar{\mu}(4)\gamma^\sigma \gamma p \gamma^\mu \frac{1-\gamma_5}{2} e(1)] \\ & + \frac{\xi - 1}{p^2 - \xi M_W^2} [\bar{\mu}(3)\gamma p \gamma p \gamma_\sigma \frac{1-\gamma_5}{2} e(2)][\bar{\mu}(4)\gamma^\sigma \gamma p \gamma p \frac{1-\gamma_5}{2} e(1)] \\ & + \frac{\xi - 1}{p^2 - \xi M_W^2} [\bar{\mu}(3)\gamma_\mu \gamma p \gamma p \frac{1-\gamma_5}{2} e(2)][\bar{\mu}(4)\gamma p \gamma p \gamma^\mu \frac{1-\gamma_5}{2} e(1)] \\ & + \frac{(\xi - 1)^2}{(p^2 - \xi M_W^2)^2} [\bar{\mu}(3)\gamma p \gamma p \gamma p \frac{1-\gamma_5}{2} e(2)][\bar{\mu}(4)\gamma p \gamma p \gamma p \frac{1-\gamma_5}{2} e(1)] \Big\} \end{aligned} \tag{B.2}$$

We list two useful equations as below:

$$\int \frac{d^4p}{(2\pi)^4} p^\mu p^\nu f(p^2) = \frac{g^{\mu\nu}}{4} \int \frac{d^4p}{(2\pi)^4} p^2 f(p^2) \tag{B.3}$$

$$\begin{aligned}&[\bar{\mu}(3)\gamma_\mu\gamma_\alpha\gamma_\nu \frac{1-\gamma_5}{2} e(2)][\bar{\mu}(4)\gamma^\nu\gamma^\alpha\gamma^\mu \frac{1-\gamma_5}{2} e(1)] \\ &= 4 \cdot [\bar{\mu}(3)\gamma_\mu \frac{1-\gamma_5}{2} e(2)][\bar{\mu}(4)\gamma^\mu \frac{1-\gamma_5}{2} e(1)]\end{aligned} \tag{B.4}$$

By using above two equations we can easily simplify the T-matrix element as

$$\begin{aligned}iT_{a1} = &\left(\frac{g}{\sqrt{2}}\right)^4 \sum_{A=1}^{6}\sum_{B=1}^{6} (V_{\mu A}V_{eA}^*)(VV_{eB}^*)[\bar{\mu}(3)\gamma_\mu \frac{1-\gamma_5}{2} e(2)][\bar{\mu}(4)\gamma^\mu \frac{1-\gamma_5}{2} e(1)] \\ &\cdot \int \frac{d^4p}{(2\pi)^4} \frac{1}{(p^2-m_A^2)(p^2-m_B^2)(p^2-M_W^2)^2} \cdot \{p^2 + \frac{2(\xi-1)}{p^2-\xi M_W^2} \cdot p^4 \\ &+ \frac{(\xi-1)^2}{4(p^2-\xi M_W^2)^2} \cdot p^6\}\end{aligned} \tag{B.5}$$

The integral in Eq. (B.1) could be evaluated by doing Wick's rotation.

$$\begin{aligned}&\int \frac{d^4p}{(2\pi)^4} \frac{1}{(p^2-m_A^2)(p^2-m_B^2)(p^2-M_W^2)^2} \\ &\quad \cdot \left\{p^2 + \frac{2(\xi-1)}{p^2-\xi M_W^2} \cdot p^4 + \frac{(\xi-1)^2}{4(p^2-\xi M_W^2)^2} \cdot p^6\right\} \\ &= -i \int \frac{d|p_E||p_E|^3\Omega_4}{(2\pi)^4} \frac{1}{(|p_E|^2+m_A^2)(|p_E|^2+m_B^2)(|p_E|^2+M_W^2)^2} \\ &\quad \cdot \left\{|p_E|^2 + \frac{2(\xi-1)}{|p_E|^2+_W^2} \cdot |p_E|^4 + \frac{(\xi-1)^2}{4(|p_E|^2+\xi M_W^2)^2} \cdot |p_E|^6\right\}\end{aligned} \tag{B.6}$$

where Ω_4 is the four-dimensional solid angle, $\Omega_4 = 2\pi^2$.

$$\text{Let} \quad \frac{|P_E|^2}{M_W^2} = t, \quad x_A = \frac{m_A^2}{M_W^2}, \quad x_B = \frac{m_B^2}{M_W^2} \tag{B.7}$$

The integral (B.6) could be written as

$$-\frac{i}{16\pi^2 M_W^2} \int_0^\infty dt \frac{1}{(t+x_A)(t+x_B)(t+1)^2} \left\{t^2 + \frac{2(\xi-1)}{t+\xi} \cdot t^3 + \frac{(\xi-1)^2}{4(t+\xi)^2} \cdot t^4\right\} \tag{B.8}$$

Note that the CKM matrix V_{aA} satisfies an identity:

$$\sum_{A=1}^{6} V_{aA} V_{bA}^{*} = \delta_{ab} \tag{B.9}$$

This identity would help to reduce the power of the integrand in Eq. (B.8) by factor 2.

We will rewrite the T-matrix element and manipulate the integrand of Eq. (B.8) into several parts. This way we could explicitly show that some parts vanish by using identity (B.9).

$$\begin{aligned}
T_{a1} &= -\frac{g^4}{64\pi^2 M_W^2}[\bar{\mu}(3)\gamma_\mu \frac{1-\gamma_5}{2} e(2)][\bar{\mu}(4)\gamma^\mu \frac{1-\gamma_5}{2} e(1)] \sum_{A=1}^{6}\sum_{B=1}^{6}(V_{\mu A}V_{eA}^{*})(V_{\mu B}V_{eB}^{*}) \\
&\cdot \int_0^\infty dt \left[\frac{(t+x_A)(t+x_B) - (t+x_A)x_B - (t+x_B)x_A + x_A x_B}{(t+x_A)(t+x_B)(t+1)^2} \cdot \right. \\
&\left. \cdot \left\{ 1 + \frac{2(\xi-1)}{t+\xi} \cdot t + \frac{(\xi-1)^2}{4(t+\xi)^2} \cdot t^2 \right\} \right] \\
&= -\frac{g^4}{64\pi^2 M_W^2}[\bar{\mu}(3)\gamma_\mu \frac{1-\gamma_5}{2} e(2)][\bar{\mu}(4)\gamma^\mu \frac{1-\gamma_5}{2} e(1)] \sum_{A=1}^{6}\sum_{B=1}^{6}(V_{\mu A}V_{eA}^{*})(V_{\mu B}V_{eB}^{*}) \\
&\cdot \int_0^\infty dt \left[\left\{ \frac{1}{(t+1)^2} - \frac{x_B}{(t+x_B)(t+1)^2} - \frac{x_A}{(t+x_A)(t+1)^2} \right. \right. \\
&\left. \left. + \frac{x_A x_B}{(t+x_A)(t+x_B)(t+1)^2} \right\} \cdot \left\{ 1 + \frac{2(\xi-1)}{t+\xi} \cdot t + \frac{(\xi-1)^2}{4(t+\xi)^2} \cdot t^2 \right\} \right]
\end{aligned} \tag{B.10}$$

Since

$$\sum_{A=1}^{6} V_{\mu A} V_{eA}^{*} = 0, \quad \text{and} \quad \sum_{B=1}^{6} V_{\mu B} V_{eB}^{*} = 0, \tag{B.11}$$

any part in Eq. (B.10) independent of x_A or x_B will vanish. We will be left with the parts which depend on both x_A and x_B.

$$\begin{aligned}
T_{a1} &= -\frac{g^4}{64\pi^2 M_W^2}[\bar{\mu}(3)\gamma_\mu \frac{1-\gamma_5}{2} e(2)][\bar{\mu}(4)\gamma^\mu \frac{1-\gamma_5}{2} e(1)] \sum_{A=1}^{6}\sum_{B=1}^{6}(V_{\mu A}V_{eA}^{*})(V_{\mu B}V_{eB}^{*}) \\
&\cdot \int_0^\infty dt \left[\frac{x_A x_B}{(t+x_A)(t+x_B)(t+1)^2} \cdot \left\{ 1 + \frac{2(\xi-1)}{t+\xi} \cdot t + \frac{(\xi-1)^2}{4(t+\xi)^2} \cdot t^2 \right\} \right]
\end{aligned} \tag{B.12}$$

B.2 T-Matrix Element of Graph (a2) in Fig. 2.2

$$
\begin{aligned}
iT_{a2} = & \int \frac{d^4p}{(2\pi)^4} \sum_{A=1}^{6} \bar{\mu}(3)\left(-\frac{ig}{\sqrt{2}}\right) V_{\mu A}\gamma_\mu \frac{1-\gamma_5}{2} \cdot \frac{+m_a}{p^2 - m_A^2} \left(-\frac{ig}{\sqrt{2}M_W}\right) V_{eA}^*(m_e \frac{1+\gamma_5}{2} \\
& - m_A \frac{1-\gamma_5}{2}) e(2) \sum_{B=1}^{6} \bar{\mu}(4)\left(-\frac{ig}{\sqrt{2}M_W}\right) V(m_\mu \frac{1-\gamma_5}{2} - m_B \frac{1+\gamma_5}{2}) \frac{+m_B}{p^2 - m_B^2} \\
& \cdot \left(-\frac{ig}{\sqrt{2}}\right) V_{eB}^* \gamma_\nu \frac{1-\gamma_5}{2} e(1) \cdot \frac{-1}{p^2 - M_W^2} [g^{\mu\nu} + \frac{(\xi - 1)p^\mu p^\nu}{p^2 - \xi M_W^2}] \cdot \frac{1}{p^2 -_W^2} \\
= & -\left(\frac{g}{\sqrt{2}}\right)^4 \frac{1}{M_W^2} \sum_{A=1}^{6} \sum_{B=1}^{6} (V V_{eA}^*)(V_{\mu B} V_{eB}^*) \\
& \cdot \int \frac{d^4p}{(2\pi)^4} \frac{1}{(p^2 - m_A^2)(p^2 - M_B^2)(p^2 - M_W^2)(p^2 -_W^2)} \\
& \cdot \left[m_e \bar{\mu}(3) \gamma^\mu \frac{1+\gamma_5}{2} e(2) - m_A^2 \bar{\mu}(3) \gamma^\mu \frac{1-\gamma_5}{2} e(2) \right] \\
& \cdot \left[m_\mu \bar{\mu}(4) \frac{1+\gamma_5}{2} \gamma_(1) - m_B^2 \bar{\mu}(4) \frac{1+\gamma_5}{2} \gamma_(1) \right] [g^{\mu\nu} + \frac{(\xi - 1)p^\mu p^\nu}{p^2 - \xi M_W^2}]
\end{aligned}
\tag{B.13}
$$

We have Dirac's equation:

$$(\gamma p - m)u(p) = 0 \tag{B.14}$$

$$\bar{u}(p)(\gamma p - m) = 0 \tag{B.15}$$

Hence,

$$m_e \bar{\mu}(3) \gamma^\mu \gamma p \frac{1+\gamma_5}{2} e(2) = \bar{\mu}(3) \gamma^\mu \gamma p \frac{1+\gamma_5}{2} \gamma p_2 e(2) \tag{B.16}$$

$$m_\mu \bar{\mu}(4) \gamma p \frac{1+\gamma_5}{2} \gamma_\nu e(1) = \bar{\mu}(4) \gamma p_4 \gamma p \frac{1+\gamma_5}{2} \gamma_\nu e(1) \tag{B.17}$$

Since we ignore all the external momenta in our discussion, the above terms will vanish. We could simplify the T-matrix element as follows:

$$
\begin{aligned}
iT_{a2} = & -\left(\frac{g}{\sqrt{2}}\right)^4 \frac{1}{M_W^2} [\bar{\mu}(3) \gamma_\mu \frac{1-\gamma_5}{2} e(2)] [\bar{\mu}(4) \gamma^\mu \frac{1-\gamma_5}{2} e(1)] \sum_{A=1}^{6} \sum_{B=1}^{6} (V_{\mu A} V_{eA}^*)(V_{\mu B} V_{eB}^*) \\
& \cdot \int \frac{d^4p}{(2\pi)^4} \frac{m_A^2 m_B^2}{(p^2 - m_A^2)(p^2 - M_B^2)(p^2 - M_W^2)(p^2 - \xi M_W^2)} [1 + \frac{\xi - 1}{4(p^2 - \xi M_W^2)} \cdot p^2].
\end{aligned}
\tag{B.18}
$$

Doing Wick's rotation and expressing the result in terms of t, x_A and x_B, we have

$$T_{a2} = -\frac{g^4}{64\pi^2 M_W^2}[\bar{\mu}(3)\gamma_\mu \frac{1-\gamma_5}{2} e(2)][\bar{\mu}(4)\gamma^\mu \frac{1-\gamma_5}{2} e(1)] \sum_{A=1}^{6}\sum_{B=1}^{6}(V_{\mu A}V_{eA}^*)(V_{\mu B}V_{eB}^*)$$
$$\cdot \int_0^\infty dt \left[\frac{x_A x_B}{(t+x_A)(t+x_B)(t+1)(t+\xi)} \cdot \left\{t + \frac{\xi - 1}{4(t+\xi)} \cdot t^2\right\}\right] \tag{B.19}$$

B.3 T-Matrix Element of Graph (a3) in Fig. 2.2

It turned out graph $(a3)$ has the same T-matrix element as $(a2)$.

B.4 T-Matrix Element of Graph (a4) in Fig. 2.2

$$iT_{a4} = \left(-\frac{ig}{\sqrt{2}M_W}\right)^4 \int \frac{d^4p}{(2\pi)^4} \sum_{A=1}^{6} \bar{\mu}(3) V_{\mu A}(m_\mu \frac{1-\gamma_5}{2} - m_A \frac{1+\gamma_5}{2}) \frac{\gamma p + m_A}{p^2 - m_A^2} V_{eA}^*$$
$$\cdot (m_e \frac{1+\gamma_5}{2} - m_A \frac{1-\gamma_5}{2}) e(2) \cdot \sum_{B=1}^{6} \bar{\mu}(4) V_{\mu B}(m_\mu \frac{1-\gamma_5}{2} - m_B \frac{1+\gamma_5}{2})$$
$$\cdot \frac{\gamma p + m_B}{p^2 - m_B^2} V_{eB}^* \cdot (m_e \frac{1+\gamma_5}{2} - m_B \frac{1-\gamma_5}{2}) e(1) \cdot \frac{1}{(p^2 - \xi M_W^2)^2} \tag{B.20}$$

Like what we did in Eqs. (B.16) and (B.17), we could ignore all the terms with m_μ or m_e in Eq. (B.20). Then we have

$$iT_{a4} = \left(\frac{g}{\sqrt{2}M_W}\right)^4 [\bar{\mu}(3)\gamma_\mu \frac{1-\gamma_5}{2} e(2)][\bar{\mu}(4)\gamma^\mu \frac{1-\gamma_5}{2} e(1)] \sum_{A=1}^{6}\sum_{B=1}^{6}(V_{\mu A}V_{eA}^*)(V_{\mu B}V_{eB}^*)$$
$$\cdot \int \frac{d^4p}{(2\pi)^4} \frac{m_A^2 m_B^2}{(p^2 - m_A^2)(p^2 - m_B^2)(p^2 - \xi M_W^2)^2} \cdot \frac{p^2}{4} \tag{B.21}$$

By doing Wick's rotation and expressing above equation in terms of t, x_A and x_B we can simplify the T-matrix element as

$$T_{a4} = -\frac{g^4}{64\pi^2 M_W^2}[\bar{\mu}(3)\gamma_\mu \frac{1-\gamma_5}{2} e(2)][\bar{\mu}(4)\gamma^\mu \frac{1-\gamma_5}{2} e(1)] \sum_{A=1}^{6}\sum_{B=1}^{6}(V_{\mu A}V_{eA}^*)(V_{\mu B}V_{eB}^*)$$
$$\cdot \int_0^\infty dt \left[\frac{x_A x_B}{(t+x_A)(t+x_B)(t+\xi)^2} \cdot \frac{t^2}{4}\right] \tag{B.22}$$

B.5 T-Matrix Element of Graph (c1) in Fig. 2.3

The T-matrix element of graph $(c1)$ is

$$iT_{c1} = -\left(-\frac{ig}{\sqrt{2}}\right)^4 \int \frac{d^4p}{(2\pi)^4} \left\{ \sum_{A=1}^{6} (V_{\mu A})^2 \bar{\mu}(3)\gamma_\mu \frac{m_A}{p^2 - m_A^2} \gamma_\rho \frac{1+\gamma_5}{2} \mu^c(4) \right.$$
$$\cdot \sum_{B=1}^{6} (V_{eB}^*)^2 \bar{e^c}(2)\gamma_\nu \frac{m_B}{p^2 - m_B^2} \gamma_\sigma \frac{1-\gamma_5}{2} e(1)$$
$$\cdot \frac{1}{p^2 - M_W^2 + i\epsilon} \cdot [g^{\mu\sigma} + \frac{(\xi - 1)p^\mu p^\sigma}{p^2 - \xi M_W^2}]$$
$$\left. \cdot \frac{1}{p^2 - M_W^2 + i\epsilon} \cdot [g^{\nu\rho} + \frac{(\xi - 1)p^\nu p^\rho}{p^2 - \xi M_W^2}] \right\} \tag{B.23}$$

We multiply out every thing and write above equation in an explicit form:

$$iT_{c1} = -\left(\frac{g}{\sqrt{2}}\right)^4 \int \frac{d^4p}{(2\pi)^4} \left\{ \sum_{A=1}^{6} \sum_{B=1}^{6} (V_{\mu A})^2 (V_{eB}^*)^2 \frac{m_A m_B}{(p^2 - m_A^2)(p^2 - m_B^2)(p^2 - M_W^2)^2} \right.$$
$$\cdot \left\{ [\bar{\mu}(3)\gamma_\mu \gamma_\rho \frac{1+\gamma_5}{2} \mu^c(4)][\bar{e^c}(2)\gamma^\rho \gamma^\mu \frac{1-\gamma_5}{2} e(1)] \right.$$
$$+ [\bar{\mu}(3)\gamma_\mu \gamma p \frac{1+\gamma_5}{2} \mu^c(4)][\bar{e^c}(2)\gamma p \gamma^\mu \frac{1-\gamma_5}{2} e(1)] \cdot \frac{\xi - 1}{p^2 - \xi M_W^2}$$
$$+ [\bar{\mu}(3)\gamma p \gamma_\rho \frac{1+\gamma_5}{2} \mu^c(4)][\bar{e^c}(2)\gamma^\rho \gamma p \frac{1-\gamma_5}{2} e(1)] \cdot \frac{\xi - 1}{p^2 - \xi M_W^2}$$
$$\left. \left. + [\bar{\mu}(3)\gamma p \gamma p \frac{1+\gamma_5}{2} \mu^c(4)][\bar{e^c}(2)\gamma p \gamma p \frac{1-\gamma_5}{2} e(1)] \cdot \left(\frac{\xi - 1}{p^2 - \xi M_W^2}\right)^2 \right\} \right\} \tag{B.24}$$

We list two useful equations as below:

$$[u_1 \gamma_\mu \gamma_\nu \frac{1+\gamma_5}{2} u_2][u_3 \gamma^\mu \gamma^\nu \frac{1-\gamma_5}{2} u_4]$$
$$= [u_1 \gamma_\mu \gamma_\nu \frac{1+\gamma_5}{2} u_2][u_3 \gamma^\nu \gamma^\mu \frac{1-\gamma_5}{2} u_4]$$
$$= 4 \cdot [u_1 \frac{1+\gamma_5}{2} u_2][u_3 \frac{1-\gamma_5}{2} u_4] \tag{B.25}$$

$$\int \frac{d^4p}{(2\pi)^4} p^\mu p^\nu f(p^2) = \frac{g^{\mu\nu}}{4} \int \frac{d^4p}{(2\pi)^4} p^2 f(p^2) \tag{B.26}$$

By using above equations we can easily simplify the T-matrix element as

$$iT_{c1} = -\left(\frac{g}{\sqrt{2}}\right)^4 \sum_{A=1}^{6}(V_{\mu A})^2 m_A \sum_{B=1}^{6}(V_{eB}^*)^2 m_B \cdot [\bar{\mu}(3)\frac{1+\gamma_5}{2}\mu^c(4)][\bar{e^c}(2)\frac{1-\gamma_5}{2}e(1)]$$
$$\cdot \int \frac{d^4p}{(2\pi)^4}\left\{\frac{1}{(p^2-m_A^2)(p^2-m_B^2)(p^2-M_W^2)^2}\cdot\left(4+\frac{2(\xi-1)p^2}{p^2-\xi M_W^2}\right.\right.$$
$$\left.\left.+\frac{(\xi-1)^2p^4}{(p^2-\xi M_W^2)^2}\right)\right\} \tag{B.27}$$

We could evaluate the integration in Eq. (B.27) by doing Wick's rotation.

$$\int \frac{d^4p}{(2\pi)^4}\left\{\frac{1}{(p^2-m_A^2)(p^2-m_B^2)(p^2-M_W^2)^2}\cdot\left(4+\frac{2(\xi-1)p^2}{p^2-\xi M_W^2}+\frac{(\xi-1)^2p^4}{(p^2-\xi M_W^2)^2}\right)\right\}$$
$$= i\int \frac{d|P_E||P_E|^3\Omega_4}{(2\pi)^4}\frac{1}{(|P_E|^2+m_A^2)(|P_E|^2+m_B^2)(|P_E|^2+M_W^2)^2}$$
$$\cdot\left(4+\frac{2(\xi-1)|P_E|^2}{|P_E|^2+\xi M_W^2}+\frac{(\xi-1)^2|P_E|^4}{(|P_E|^2+\xi M_W^2)^2}\right) \tag{B.28}$$

where Ω_4 is the four-dimensional solid angle, $\Omega_4 = 2\pi^2$.

$$\text{Let}\quad \frac{|P_E|^2}{M_W^2} = t,\quad x_A = \frac{m_A^2}{M_W^2},\quad x_B = \frac{m_B^2}{M_W^2} \tag{B.29}$$

The integral (B.28) could be written as

$$\frac{i}{16\pi^2 M_W^4}\int_0^\infty dt\left\{\frac{t}{(t+x_A)(t+x_B)(t+1)^2}\cdot\left(4+\frac{2(\xi-1)t}{t+\xi}+\frac{(\xi-1)^2t^2}{(t+\xi)^2}\right)\right\} \tag{B.30}$$

Hence, we could eventually write the T-matrix element of graph $(c1)$ as

$$T_{c1} = -\frac{g^4}{64\pi^2 M_W^4}[\bar{\mu}(3)\frac{1+\gamma_5}{2}\mu^c(4)][\bar{e^c}(2)\frac{1-\gamma_5}{2}e(1)]\sum_{A=1}^{6}(V_{\mu A})^2 m_A\sum_{B=1}^{6}(V_{eB}^*)^2 m_B$$
$$\cdot\int_0^\infty dt\left\{\frac{t}{(t+x_A)(t+x_B)(t+1)^2}\cdot\left(4+\frac{2(\xi-1)t}{t+\xi}+\frac{(\xi-1)^2t^2}{(t+\xi)^2}\right)\right\} \tag{B.31}$$

Note that the CKM matrix V_{aA} satisfies an identity (Ilakovac et al. 1995):

$$\sum_{k}^{6} V_{lk} V_{l'k} m_k = 0, \tag{B.32}$$

which is forced by the properties of the neutrino mass matrix of seesaw model. In seesaw model the generation of left-handed Majorana neutrino masses requires a mass dimension-five operator. Since the dimension-five operator is non-renormalizable, we would like to suppress this mass part in seesaw model. This would generate the identity (B.32). We will show that this identity would help to reduce the power of the integrand in Eq. (B.31) by factor 2.

We will rewrite the T-matrix element of graph (c1) and manipulate the integrand of Eq. (B.31) into several parts. This way we could explicitly show that some parts just vanish by using identity (B.32).

$$\begin{aligned}
T_{c1} = &-\frac{g^4}{64\pi^2 M_W^4}[\bar{\mu}(3)\frac{1+\gamma_5}{2}\mu^c(4)][\bar{e^c}(2)\frac{1-\gamma_5}{2}e(1)]\sum_{A=1}^{6}(V_{\mu A})^2 m_A \sum_{B=1}^{6}(V_{eB}^*)^2 m_B \\
&\cdot \int_0^\infty dt \Bigg[\frac{4t}{(t+x_A)(t+x_B)(t+1)^2} \\
&+2(\xi-1)\cdot\frac{(t+x_A)(t+x_B)-(t+x_A)x_B-(t+x_B)x_A+x_A x_B}{(t+x_A)(t+x_B)(t+1)^2(t+\xi)} \\
&+(\xi-1)^2\cdot\frac{t(t+x_A)(t+x_B)-t(t+x_A)x_B-t(t+x_B)x_A+x_A x_B t}{(t+x_A)(t+x_B)(t+1)^2(t+\xi)^2}\Bigg] \\
= &-\frac{g^4}{64\pi^2 M_W^4}[\bar{\mu}(3)\frac{1+\gamma_5}{2}\mu^c(4)][\bar{e^c}(2)\frac{1-\gamma_5}{2}e(1)]\sum_{A=1}^{6}(V_{\mu A})^2 m_A \sum_{B=1}^{6}(V_{eB}^*)^2 m_B \\
&\cdot \int_0^\infty dt \Bigg[\frac{4t}{(t+x_A)(t+x_B)(t+1)^2}+2(\xi-1)\cdot\Bigg(\frac{1}{(t+1)^2(t+\xi)} \\
&-\frac{x_B}{(t+x_B)(t+1)^2(t+\xi)}-\frac{x_A}{(t+x_A)(t+1)^2(t+\xi)} \\
&+\frac{x_A x_B}{(t+x_A)(t+x_B)(t+1)^2(t+\xi)}\Bigg)+(\xi-1)^2\cdot\Bigg(\frac{t}{(t+1)^2(t+\xi)^2} \\
&-\frac{x_B t}{(t+x_B)(t+1)^2(t+\xi)^2}-\frac{x_A t}{(t+x_A)(t+1)^2(t+\xi)^2} \\
&+\frac{x_A x_B t}{(t+x_A)(t+x_B)(t+1)^2(t+\xi)^2}\Bigg)\Bigg]
\end{aligned} \tag{B.33}$$

We have

$$\sum_{A=1}^{6}(V_{\mu A})^2 m_A = 0, \quad \text{and} \quad \sum_{B=1}^{6}(V_{eB}^*)^2 m_B = 0 \tag{B.34}$$

Hence, any part of the integrand independent of x_A or x_B will vanish. We will be left with the parts which depend on both x_A and x_B.

$$\begin{aligned} T_{c1} = & -\frac{g^4}{64\pi^2 M_W^4}[\bar{\mu}(3)\frac{1+\gamma_5}{2}\mu^c(4)][\bar{e^c}(2)\frac{1-\gamma_5}{2}e(1)]\sum_{A=1}^{6}(V_{\mu A})^2 m_A \sum_{B=1}^{6}(V_{eB}^*)^2 m_B \\ & \cdot \int_0^\infty dt \left[\frac{4t}{(t+x_A)(t+x_B)(t+1)^2} + \frac{2(\xi-1)x_A x_B}{(t+x_A)(t+x_B)(t+1)^2(t+\xi)} \right. \\ & \left. + \frac{(\xi-1)^2 x_A x_B \cdot t}{(t+x_A)(t+x_B)(t+1)^2(t+\xi)^2}\right] \end{aligned} \tag{B.35}$$

B.6 T-Matrix Element of Graph (c2) in Fig. 2.3

$$\begin{aligned} iT_{c2} = & \left(\frac{g}{\sqrt{2}}\right)^4 \frac{1}{M_W^2}\sum_{A=1}^{6}\sum_{B=1}^{6}(V_{\mu A})^2(V_{eB}^*)^2 \\ & \cdot \int \frac{d^4p}{(2\pi)^4}\left\{\left[\frac{m_\mu m_A}{p^2-m_A^2}\bar{\mu}(3)\frac{1-\gamma_5}{2}\gamma_\rho \mu^c(4) - \frac{m_A}{p^2-m_A^2}\bar{\mu}(3)\gamma p\frac{1-\gamma_5}{2}\gamma_\rho \mu^c(4)\right]\right. \\ & \cdot \left[\frac{m_e m_B}{p^2-m_B^2}\bar{e^c}(2)\gamma^\nu\frac{1+\gamma_5}{2}e(1) - \frac{m_b}{p^2-m_B^2}\bar{e^c}(2)\gamma^\nu\gamma p\frac{1-\gamma_5}{2}e(1)\right] \\ & \left. \cdot \frac{1}{p^2-M_W^2}[g^{\rho\nu} + \frac{(\xi-1)p^\rho p^\nu}{p^2-\xi M_W^2}] \cdot \frac{1}{p^2-\xi M_W^2}\right\} \end{aligned} \tag{B.36}$$

Multiply every thing out, we have

$$\begin{aligned} iT_{c2} = & \left(\frac{g}{\sqrt{2}}\right)^4 \frac{1}{M_W^2}\sum_{A=1}^{6}\sum_{B=1}^{6}(V_{\mu A})^2(V_{eB}^*)^2 m_A m_B \\ & \cdot \int \frac{d^4p}{(2\pi)^4}\left\{\frac{1}{(p^2-m_A^2)(p^2-m_B^2)(p^2-M_W^2)(p^2-\xi M_W^2)}\right. \\ & \cdot \left(m_\mu m_e[\bar{\mu}(3)\frac{1-\gamma_5}{2}\gamma_\rho \mu^c(4)][\bar{e^c}(2)\gamma^\rho\frac{1+\gamma_5}{2}e(1)]\right. \\ & + \frac{m_\mu m_e(\xi-1)}{p^2-\xi M_W^2}[\bar{\mu}(3)\frac{1-\gamma_5}{2}\gamma p \mu^c(4)][\bar{e^c}(2)\gamma p\frac{1+\gamma_5}{2}e(1)] \\ & - m_\mu[\bar{\mu}(3)\frac{1-\gamma_5}{2}\gamma_\rho \mu^c(4)][\bar{e^c}(2)\gamma^\rho \gamma p\frac{1-\gamma_5}{2}e(1)] \end{aligned}$$

$$
\begin{aligned}
&-\frac{m_\mu(\xi-1)}{p^2-\xi M_W^2}[\bar{\mu}(3)\frac{1-\gamma_5}{2}\gamma p\mu^c(4)][\bar{e^c}(2)\gamma p\gamma p\frac{1-\gamma_5}{2}e(1)]\\
&-m_e[\bar{\mu}(3)\gamma p\frac{1-\gamma_5}{2}\gamma_\rho\mu^c(4)][\bar{e^c}(2)\gamma^\rho\frac{1+\gamma_5}{2}e(1)]\\
&-\frac{m_e(\xi-1)}{p^2-\xi M_W^2}[\bar{\mu}(3)\gamma p\frac{1-\gamma_5}{2}\gamma p\mu^c(4)][\bar{e^c}(2)\gamma p\frac{1+\gamma_5}{2}e(1)]\\
&+[\bar{\mu}(3)\gamma p\frac{1-\gamma_5}{2}\gamma_\rho\mu^c(4)][\bar{e^c}(2)\gamma^\rho\gamma p\frac{1-\gamma_5}{2}e(1)]\\
&+\frac{(\xi-1)}{p^2-\xi M_W^2}[\bar{\mu}(3)\gamma p\frac{1-\gamma_5}{2}\gamma p\mu^c(4)][\bar{e^c}(2)\gamma p\gamma p\frac{1-\gamma_5}{2}e(1)]\Bigg)\Bigg\}
\end{aligned}
\tag{B.37}
$$

We have Dirac's equation:

$$(\gamma p-m)u(p)=0 \tag{B.38}$$

$$\bar{u}(p)(\gamma p-m)=0 \tag{B.39}$$

Hence,

$$m_\mu[\bar{\mu}(3)\frac{1-\gamma_5}{2}\gamma_\rho\mu^c(4)]=[\bar{\mu}(3)\gamma p_3\frac{1-\gamma_5}{2}\gamma_\rho\mu^c(4)] \tag{B.40}$$

$$m_e[\bar{e^c}(2)\gamma^\rho\frac{1+\gamma_5}{2}e(1)]=[\bar{e^c}(2)\gamma^\rho\frac{1+\gamma_5}{2}\gamma p_1 e(1)] \tag{B.41}$$

Since we ignore all the external momenta in our discussion, the above terms will vanish. By the same token, we may neglect all the terms with m_μ or m_e in Eq. (B.37). Then, we have

$$
\begin{aligned}
iT_{c2}=&\left(\frac{g}{\sqrt{2}}\right)^4\frac{1}{M_W^2}\sum_{A=1}^{6}\sum_{B=1}^{6}(V_{\mu A})^2(V_{eB}^*)^2 m_A m_B\\
&\cdot\int\frac{d^4p}{(2\pi)^4}\Bigg\{\frac{1}{(p^2-m_A^2)(p^2-m_B^2)(p^2-M_W^2)(p^2-\xi M_W^2)}\\
&\cdot\Bigg([\bar{\mu}(3)\gamma p\frac{1-\gamma_5}{2}\gamma_\rho\mu^c(4)][\bar{e^c}(2)\gamma^\rho\gamma p\frac{1-\gamma_5}{2}e(1)]\\
&+\frac{(\xi-1)}{p^2-\xi M_W^2}[\bar{\mu}(3)\gamma p\frac{1-\gamma_5}{2}\gamma p\mu^c(4)][\bar{e^c}(2)\gamma p\gamma p\frac{1-\gamma_5}{2}e(1)]\Bigg)\Bigg\}
\end{aligned}
\tag{B.42}
$$

And it could be simplified further by Eqs. (B.25) and (B.26) as

$$
\begin{aligned}
iT_{c2}=&\left(\frac{g}{\sqrt{2}}\right)^4\frac{1}{M_W^2}[\bar{\mu}(3)\frac{1+\gamma_5}{2}\mu^c(4)][\bar{e^c}(2)\frac{1-\gamma_5}{2}e(1)]\sum_{A=1}^{6}\sum_{B=1}^{6}(V_{\mu A})^2(V_{eB}^*)^2 m_A m_B\\
&\cdot\int\frac{d^4p}{(2\pi)^4}\Bigg\{\frac{1}{(p^2-m_A^2)(p^2-m_B^2)(p^2-M_W^2)(p^2-\xi M_W^2)}\cdot\left(p^2+\frac{(\xi-1)}{p^2-\xi M_W^2}\right)\Bigg\}
\end{aligned}
\tag{B.43}
$$

As what we did for graph $(c1)$ we can also write the T-matrix element in terms of t, x_A and x_B.

$$T_{c2} = -\frac{g^4}{64\pi^2 M_W^4}[\bar{\mu}(3)\frac{1+\gamma_5}{2}\mu^c(4)][\bar{e^c}(2)\frac{1-\gamma_5}{2}e(1)]\sum_{A=1}^{6}(V_{\mu A})^2 m_A \sum_{B=1}^{6}(V_{eB}^*)^2 m_B$$
$$\cdot \int_0^\infty dt \left[\frac{t}{(t+x_A)(t+x_B)(t+1)(t+\xi)}\cdot\left(t+\frac{(\xi-1)t^2}{t+\xi}\right)\right] \quad \text{(B.44)}$$

Using Eq. (B.32) to get rid of the vanishing parts,

$$T_{c2} = -\frac{g^4}{64\pi^2 M_W^4}[\bar{\mu}(3)\frac{1+\gamma_5}{2}\mu^c(4)][\bar{e^c}(2)\frac{1-\gamma_5}{2}e(1)]\sum_{A=1}^{6}(V_{\mu A})^2 m_A \sum_{B=1}^{6}(V_{eB}^*)^2 m_B$$
$$\cdot \int_0^\infty dt \left[\frac{x_A x_B}{(t+x_A)(t+x_B)(t+1)(t+\xi)} + \frac{(\xi-1)x_A x_B \cdot t}{(t+x_A)(t+x_B)(t+1)(t+\xi)^2}\right] \quad \text{(B.45)}$$

B.7 T-Matrix Element of Graph (c3) in Fig. 2.3

It turned out graph $(c3)$ has the same T-matrix element as $(c2)$.

B.8 T-Matrix element of graph (c4) in Fig. 2.3

$$iT_{c4} = -\left(\frac{-ig}{\sqrt{2}M_W}\right)^4 \sum_{A=1}^{6}\sum_{B=1}^{6}(V_{\mu A})^2(V_{eB}^*)^2$$
$$\cdot \int \frac{d^4}{(2\pi)^4}\left\{\left(-\frac{m_\mu^2 m_A}{p^2-m_A^2}[\bar{\mu}(3)\frac{1-\gamma_5}{2}\mu^c(4)] + \frac{m_\mu m_A}{p^2-m_A^2}[\bar{\mu}(3)\gamma p\frac{1+\gamma_5}{2}\mu^c(4)]\right.\right.$$
$$+\frac{m_\mu m_A}{p^2-m_A^2}[\bar{\mu}(3)\gamma p\frac{1-\gamma_5}{2}\mu^c(4)] - \frac{m_A^3}{p^2-m_A^2}[\bar{\mu}(3)\frac{1+\gamma_5}{2}\mu^c(4)]\Bigg)$$
$$\cdot\Bigg(-\frac{m_e^2 m_B}{p^2-m_B^2}[\bar{e^c}(2)\frac{1+\gamma_5}{2}e(1)] + \frac{m_e m_B}{p^2-m_B^2}[\bar{e^c}(2)\gamma p\frac{1-\gamma_5}{2}e(1)]$$
$$\left.\cdot\frac{m_B m_e}{p^2-m_B^2}[\bar{e^c}(2)\gamma p\frac{1+\gamma_5}{2}e(1)] - \frac{m_B^3}{p^2-m_B^2}[\bar{e^c}(2)\frac{1-\gamma_5}{2}e(1)]\Bigg)\cdot(\frac{1}{p^2-\xi M_W^2})^2\right\} \quad \text{(B.46)}$$

Ignoring the external momenta, we can simplify above equation as

$$iT_{c4} = -\left(\frac{-ig}{\sqrt{2}M_W}\right)^4 [\bar{\mu}(3)\frac{1+\gamma_5}{2}\mu^c(4)][\bar{e^c}(2)\frac{1-\gamma_5}{2}e(1)]\sum_{A=1}^{6}\sum_{B=1}^{6}(V_{\mu A})^2(V_{eB}^*)^2$$
$$\cdot \int \frac{d^4}{(2\pi)^4}\frac{m_A^3 m_B^3}{(p^2-m_A^2)(p^2-m_B^2)(p^2-\xi M_W^2)^2} \quad \text{(B.47)}$$

Doing Wick'rotation and expressing the T-matrix element in terms of t, m_A and m_B, we have

$$T_{c4} = -\frac{g^4}{64\pi^2 M_W^4}[\bar{\mu}(3)\frac{1+\gamma_5}{2}\mu^c(4)][\bar{e^c}(2)\frac{1-\gamma_5}{2}e(1)]\sum_{A=1}^{6}(V_{\mu A})^2 m_A \sum_{B=1}^{6}(V_{eB}^*)^2 m_B$$
$$\cdot \int_0^\infty dt \frac{x_A x_B \cdot t}{(t+x_A)(t+x_B)(t+\xi)^2} \quad \text{(B.48)}$$

Appendix C. General Formalism of Muonium–Anitmuonium Time Evolution

Muonium (antimuonium) is a non-relativistic Coulombic bound state of an electron and an anti-muon (positron and muon). $|M\rangle$ (muonium) and $|\bar{M}\rangle$ (antimuonium) are mixed by the weak interaction. Thus, in the presence of this type of interaction, the two states are inseparable; they form a basis for a two-dimensional subspace. We may write,

$$\frac{1}{\sqrt{\langle M|M\rangle}}|M\rangle = \begin{pmatrix}1\\0\end{pmatrix}, \quad \frac{1}{\sqrt{\langle \bar{M}|\bar{M}\rangle}}|\bar{M}\rangle = \begin{pmatrix}0\\1\end{pmatrix} \quad \text{(C.1)}$$

Let $|\psi(t)\rangle$ be an arbitrary state of such space,

$$|\psi(t)\rangle = A(t)\frac{1}{\sqrt{\langle M|M\rangle}}|M\rangle + B(t)\frac{1}{\sqrt{\langle \bar{M}|\bar{M}\rangle}}|\bar{M}\rangle = \begin{pmatrix}A(t)\\B(t)\end{pmatrix} \quad \text{(C.2)}$$

The time evolution of $\psi(t)$ is described by

$$i\frac{d}{dt}\psi(t) = H\psi(t) \quad \text{(C.3)}$$

where

$$\frac{\langle i|H|j\rangle}{\sqrt{\langle i|i\rangle\langle j|j\rangle}} = M_{ij} - \frac{i}{2}\Gamma_{ij} \quad \text{(C.4)}$$

In our two-dimensional space H has the matrix representation

$$H = M - \frac{i}{2}\Gamma = \begin{pmatrix} M_{11} - \frac{i}{2}\Gamma_{11} & M_{12} - \frac{i}{2}\Gamma_{12} \\ M_{21} - \frac{i}{2}\Gamma_{21} & M_{22} - \frac{i}{2}\Gamma_{22} \end{pmatrix} \tag{C.5}$$

CPT invariance of H implies that $M_{11} = M_{22} \equiv M_0$ and $\Gamma_{11} = \Gamma_{22} \equiv \Gamma_0$. Since by construction, both M and Γ are Hermitian, M_0 and Γ_0 are both real numbers, and $M_{21} = M_{12}^*$ and $\Gamma_{21} = \Gamma_{12}^*$. Hence

$$H = \begin{pmatrix} M_0 - \frac{i}{2}\Gamma_0 & M_{12} - \frac{i}{2}\Gamma_{12} \\ M_{12}^* - \frac{i}{2}\Gamma_{12}^* & M_0 - \frac{i}{2}\Gamma_0 \end{pmatrix} \tag{C.6}$$

Diagonalizing (C.6), we find two mass eigen states:

$$|M_\pm\rangle = \frac{1}{\sqrt{2(1+|\varepsilon|^2)}}[(1+\varepsilon)|M\rangle \pm (1-\varepsilon)|\bar{M}\rangle] \tag{C.7}$$

where

$$\varepsilon = \frac{\sqrt{\mathcal{M}_{M\bar{M}}} - \sqrt{\mathcal{M}_{\bar{M}M}}}{\sqrt{\mathcal{M}_{M\bar{M}}} + \sqrt{\mathcal{M}_{\bar{M}M}}} \tag{C.8}$$

$$\mathcal{M}_{M\bar{M}} = \frac{\langle M| - \int d^3r \mathcal{L}_{\rm eff}|\bar{M}\rangle}{\sqrt{\langle M|M\rangle\langle\bar{M}|\bar{M}\rangle}}, \quad \mathcal{M}_{\bar{M}M} = \frac{\langle\bar{M}| - \int d^3r \mathcal{L}_{\rm eff}|M\rangle}{\sqrt{\langle M|M\rangle\langle\bar{M}|\bar{M}\rangle}} \tag{C.9}$$

The magnitude of the mass splitting between the two mass eigenstates is $|\Delta M|$

$$\begin{aligned} M_+ - \frac{i}{2}\Gamma_+ &= (M_0 - \frac{i}{2}\Gamma_0) + \sqrt{(M_{12} - \frac{i}{2}\Gamma_{12})(M_{12}^* - \frac{i}{2}\Gamma_{12}^*)} \\ M_- - \frac{i}{2}\Gamma_- &= (M_0 - \frac{i}{2}\Gamma_0) - \sqrt{(M_{12} - \frac{i}{2}\Gamma_{12})(M_{12}^* - \frac{i}{2}\Gamma_{12}^*)} \end{aligned} \tag{C.10}$$

Taking a difference of above two eigenvalues

$$(M_+ - M_-) - \frac{i}{2}(\Gamma_+ - \Gamma_-) = 2\sqrt{(M_{12} - \frac{i}{2}\Gamma_{12})(M_{12}^* - \frac{i}{2}\Gamma_{12}^*)} \tag{C.11}$$

Then we have

$$\begin{aligned} |\Delta M| &= 2\left|{\rm Re}\sqrt{(M_{12} - \frac{i}{2}\Gamma_{12})(M_{12}^* - \frac{i}{2}\Gamma_{12}^*)}\right| \\ &= 2\left|{\rm Re}\sqrt{\mathcal{M}_{M\bar{M}}\mathcal{M}_{\bar{M}M}}\right| \end{aligned} \tag{C.12}$$

Since muonium and antimuonium are linear combinations of the mass diagonal states, an initially prepared muonium or antimuonium state will undergo oscillations into one another as a function of time. The oscillation time is given by $\tau_{\bar{M}M}$, and we will see this oscillation time is related to $|\Delta M|$ as $\tau_{\bar{M}M} = \frac{1}{|\Delta M|}$. According to (C.7), we have

$$
\begin{aligned}
|M\rangle &= \frac{\sqrt{2(1+|\varepsilon|^2)}}{2(1+\varepsilon)}(|M_+\rangle + |M_-\rangle) \\
|\bar{M}\rangle &= \frac{\sqrt{2(1+|\varepsilon|^2)}}{2(1-\varepsilon)}(|M_+\rangle - |M_-\rangle)
\end{aligned}
\tag{C.13}
$$

Hence, $|M\rangle$ and $|\bar{M}\rangle$ could be written in time evolution forms:

$$
\begin{aligned}
|M(t)\rangle &= \frac{\sqrt{2(1+|\varepsilon|^2)}}{2(1+\varepsilon)}\left(e^{-i(M_+-\frac{i}{2}\Gamma_+)t}|M_+(0)\rangle + e^{-i(M_--\frac{i}{2}\Gamma_-)t}|M_-(0)\rangle\right) \\
|\bar{M}(t)\rangle &= \frac{\sqrt{2(1+|\varepsilon|^2)}}{2(1-\varepsilon)}\left(e^{-i(M_+-\frac{i}{2}\Gamma_+)t}|M_+(0)\rangle - e^{-i(M_--\frac{i}{2}\Gamma_-)t}|M_-(0)\rangle\right)
\end{aligned}
\tag{C.14}
$$

The beam intensity for M and $\bar{M}$ are given by

$$
\begin{aligned}
|\langle M(t)|M(t)\rangle| = &\frac{(1+|\varepsilon|^2)}{2(1+\varepsilon)^2}\left(|\langle M_+(0)|M_+(0)\rangle|e^{-\Gamma_+t} + |\langle M_-(0)|M_-(0)\rangle|e^{-\Gamma_-t}\right. \\
&\left. + 2e^{-\frac{1}{2}(\Gamma_++\Gamma_-)t}|_+(0)|M_-(0)\rangle|\cos(\Delta Mt)\right)
\end{aligned}
\tag{C.15}
$$

and

$$
\begin{aligned}
|\langle \bar{M}(t)|\bar{M}(t)\rangle| = &\frac{(1+|\varepsilon|^2)}{2(1-\varepsilon)^2}\left(|\langle M_+(0)|M_+(0)\rangle|e^{-\Gamma_+t} + |\langle M_-(0)|M_-(0)\rangle|e^{-\Gamma_-t}\right. \\
&\left. - 2e^{-\frac{1}{2}(\Gamma_++\Gamma_-)t}|\langle M_+(0)|M_-(0)\rangle|\cos(\Delta Mt)\right)
\end{aligned}
\tag{C.16}
$$

The period of the oscillation is

$$
T = \frac{2\pi}{|\Delta M|} \tag{C.17}
$$

Approximately the oscillation time scale is

$$
\tau_{\bar{M}M} \simeq \frac{1}{|\Delta M|} \tag{C.18}
$$

Author Biography

Boyang Liu received his B.S. degree in Physics from the University of Science and Technology of China (USTC), Hefei, China, in 2003 and received his Ph.D. degree in physics from Purdue University, West Lafayette, IN, USA in 2010. He is currently working as a postdoc in the Institute of Physics, Chinese Academy of Sciences (CAS). His research interests include phenomenology of physics beyond the Standard Model, especially the supersymmetric extensions of the Standard Model and Brane World models, as well as neutrino physics, with an emphasis on models of neutrino masses, neutrino mixing and CP violation.

Reference

A. Ilakovac, A. Pilaftsis, Nucl. Phys. **B437** (1995) 491